# TEACHING CHILDREN IN THE LABORATORY

# Teaching Children in the Laboratory

JOAN SOLOMON



CROOM HELM LONDON

Croom Helm Ltd, 2-10 St John's Road, London SW11

British Library Cataloguing in Publication Data

Solomon, Joan
Teaching children in the laboratory.
1. Science – Study and teaching
I. Title
507'.1 LB1585

ISBN 0-7099-2304-X
ISBN 0-7099-2305-8 Pbk



Printed in Great Britain by Biddles Ltd, Guildford, Surrey

CONTENTS

# PREFACE

Teaching, like being an understanding parent, is an empirical art which is only made possible by listening to the spontaneous and outspoken comments of children. It was such a response that finally made the writing of this book seem urgent and important. It happened at the end of one of those 'discovery' sessions, when a suggested experiment had yielded surprising results which my pupils had then immediately followed up with tests of their own. The topic was electromagnetic induction; given a solenoid, a magnet and a galvanometer, the class had shown that although a stationary magnet did not produce a flow of current, a moving magnet did. They then, on their own initiative, collected data on the various factors that affected the magnitude of the current and it had all seemed to present an almost perfect example of the guided discovery method at work. Yet there was none of that feeling of elation which such an experimental success should have generated. Instead a girl turned to me at the end of the lesson with outright disappointment and reproach, and said, 'I can see all that but I don't understand it one bit! You haven't explained it.' Children's comments can be difficult to interpret. Was it I, rather than experiment, that was expected to 'explain' nature? Why was the experiment so unsatisfying? Isn't it true that seeing is believing? Obviously it was not. The movement of the electrons in the wire remained incomprehensible and indeed unacceptable until I had contrived by word and gesture to suggest how it might *feel* to be an electron in such a situation. Only then did the girl relax and smile; the experimental result was now meaningful and she was ready for more.

This incident, and many like it, gave me the clues I needed to unravel the curious patterns of success and failure which attended practical work. *Imaginative understanding was not a sequel to successful experiment; on the contrary, it was an essential prerequisite.* The point arose again and again in practical lessons and with its help I was able to pre-empt many a halting question and add real impact to a wide variety of experiments. Its implications were important.

In the first place, and most obviously, it offered an explanation for the obstinate difficulties which arose continually in the discovery method of teaching. In spite of its appeal to any sympathetic teacher, its practice was often very disappointing. It also seemed curious that my own children, who would beset me with questions about science and listen with fascination to tales of discovery and hypothesis, showed the same frustration and uncertainty when faced with the use of discovery as a method of teaching.

Even a reflection on the history of the great seminal discoveries of science does not lend support to this practice. From the records it becomes quite clear that it was always the initial act of imagination which both prompted the experiment and provided the framework within which it could be understood. In the words of T. S. Kuhn, one of the greatest living historians of science, 'surveying the experimental literature. . .makes one suspect that something like a paradigm is prerequisite to perception itself.' Kuhn uses the word *paradigm* to denote the established view of a particular topic, as laid out in the textbooks of the period. It includes the discoveries, the theory arising from them and the experimental procedures to which it leads. Because it pervades their education it is strongly entrenched in the thinking of scientists and is applied, consciously or unconsciously, in all their research. From there it reaches out to society at large: 'the electron is a particle' and 'cancer is a disease' are widespread and potent paradigms. If some inkling of every new revolutionary paradigm predated and prompted experiment instead of being discovered as a result of it, there can be no grounds for expecting that naïve inquiring experiment within school lessons will demonstrate the appeal and method of scientific research.

Once we accept that prepared imagination holds such a key position in the learning of science and the understanding of experiment, then our methods of teaching are bound to take account of it. The prime motives of almost every lesson will be the understanding of children's imaginative processes and the stimulation of them along scientific lines. This is an *active* philosophy of teaching, probably quite hopelessly at odds with the practice of solitary learning from a worksheet of practical instructions which has become such a feature of the last ten years of educational theory. It takes a really human teacher, not a programmed learning aid, to listen to children's imaginative comments and to recognise their imagery.[1] Only then can the teaching begin.

Although there have been many *learning* theories based upon animal experiments – rats in mazes, pigeons pecking at discs or apes reaching for bananas – established theories for the *teaching* of human children are non-existent. Certainly the contemporary studies of classroom techniques being carried out both in this country and in America which try to evaluate such factors as 'eye-ball action' and 'wait-time' between the asking of a question and the receiving of an answer do not begin to fill this gap.

The essence of teaching lies in the familiar interactions between the class and its teacher, too individual, diverse and empathic to be captured easily by any research programme. The trust and confidence it engenders, however, are of paramount importance for the cultivation of the images which will be needed in the school laboratory, all the way from the choosing of apparatus to the elucidation of scientific concepts. To lay down some new technique of teaching, even one based on the discussion of imagery, could do violence to the personal variations which are inherent in teaching. It has therefore seemed better to relate the laboratory activities being described to the practice of teaching by an illustrative anecdotal structure rather than by any formal theory.

For student and new teacher it is to be hoped that this will present a realistic picture, though it would be false and damaging to ignore the many pitfalls that obstruct the path to such a fruitful teaching dialogue. These obstacles are excavated by the kind of pupil behaviour that our societal attitudes have brought about and they can be very hard to face. Colleges of education do sterling work preparing their students, but such novices can do with all the discussion of matching practical work to behaviour problems that they can get; so a chapter on this has also been included.

This book has been written at a time when the whole profession is buzzing with new projects at almost every level. I have made reference to the better-known science syllabuses in the text whenever I have had personal knowledge of their use and impact, but the initiative in the design of syllabuses has already passed out of the hands of the large project committees to small groups or individual schools. The new format is no longer the book but the worksheet, often both written and duplicated within the school itself. These are used not only for the first two or three years, but right up to CSE, where the present total of school-devised science courses numbered more than 30,000 in 1979. Some have reacted with alarm at such proliferation, but it is a sure sign of the vitality and inventiveness of the teaching profession. The art of designing such worksheets has been learnt through trial and error on the job by these same science teachers. For this reason a chapter on the implications of worksheet learning and of the art of programmed learning from which it may be derived is to be found in the first part of the book. Worksheets not only vary in quality, but they can also serve a wide variety of purposes. The whole craft of teaching has been subtly transformed by their use, and discussion is vital.

Finally I must admit to having ignored, quite shamelessly, the whole question of course content. Readers will find no references here to such important matters as the teaching of health and sex education, environmental science or microelectronics for everyone. I intended no disparagement by concentrating on 'core' items of school science, since children learn and experiment in every practical department of science by application of the same vivid powers of growing imagination. Nor have I written one word about teaching the social implications of science, although I value most highly the sense of compassion and fierce argument for social justice that our pupils can bring to such discussion. New courses devoted to the social context of science[2] for schools are just beginning to emerge, but it might well have needed another whole book to study and describe the child's growing powers of social commitment and evaluation. A full-time teacher, of course, can always plead lack of time, even while knowing how loath she would be to mend such shortage by retreat from the boisterous, busy, but magical world of children in the laboratory.

## Notes

1. Comment from a 12-year-old boy on seeing bromine vapour diffusing into air: 'Look, the colour's coming out of it'. . .Like dye out of a dripping jersey?
2. See, for example, the SISCON-in-Schools Project.

# 1 THE LABORATORY COMES OF AGE

Science teaching must take place in a laboratory; about that at least there is no controversy. Science simply belongs there as naturally as cooking belongs in a kitchen and gardening in a garden. Books of recipes or gardening manuals can be read anywhere, but the smells, taste, labour and atmosphere can only be evoked in those who already know the reality. It is the same with science, and so the teaching of it must involve real contact with those aspects of nature which are to be studied. This has one great advantage; for young children the first move out of the classroom with its familiar rows of desks into the exotic atmosphere of a laboratory is immensely stimulating. Excitement and expectation stir. What should we do with these invaluable commodities?

The children themselves have no doubt about the answer, they are all agog to 'do experiments', to touch and use the strange new instruments and perform exciting tricks with them. To put it at its lowest level, this is a wonderful opportunity for new play! Alone amid this throng the adult teacher has another, sterner ambition – that somehow out of this rough, untaught expectancy can be forged an understanding of the insights and the methods of science. Both teacher and pupil are united only in a belief that experiment is the right tool for their objective, and indeed it is a marvellously rich instrument, if well used, with a host of possibilities. The children bring to it their gifts of imagination, curiosity and argumentativeness; the teacher hopes to add scientific purpose, logical method and practical skill. From a marriage of these two sets of qualities may issue an understanding of scientific theory.

## Early Science Teaching

Now theory is not an alluring concept in the educational scene. It conjures up memories of rigidly formulated statements, dry as chalk dust, emphatically underlined on the blackboard, to be learnt and reproduced by rote. Of course it can well be argued that these theories, the end product of so many years of scientific endeavour, are the intellectual heritage of our children. The atomic theory, the laws of motion, evolution, gravitation, electromagnetic theory, the concepts of energy and entropy, these were all hard-won laurels in the continued battle of science, giving to modern man rich and subtle ways of visualising the world around him. In the days when scientific education was still in its infancy the objective was simply to present these theories, along with their proofs, to successive generations of pupils as efficiently as possible. Here, they said in effect, is the scientific achievement, learn this.

> It may be said that a boy takes more interest in the matter by seeing for himself, or by performing for himself. . .this we admit, while we continue to doubt the educational value of the transaction. The boy would also probably take much more interest in foot-ball than in Latin grammar; but the measure of his interest is not identical with the importance of the subjects. It may be said that the fact makes a stronger impression on the boy through the medium of his sight, that he believes it the more confidently. I say that this ought not to be the case. If he does not believe the statements of his tutor – probably a clergyman of mature knowledge, recognised ability, and blameless character – his suspicion is irrational, and manifests a want of the power of appreciating evidence.[1]

That extract, it must be admitted, has been included for sheer amusement. Colonel Blimp attitudes are always good for a laugh and no subject which can be as exuberant or as catastrophic as laboratory work should be viewed with exclusive sobriety! At this period science degree courses at the universities were still very new and school science was only just beginning. Some schools and technical colleges did offer a little student practical work even in those early days; at the City of London School, for example, pupils were asked to pay an extra seven shillings a term for such experimental tuition. [2]

Gradually school laboratories equipped for real experimental work became more common and yet, at heart, the method used to teach science differed very little from that used in history, geography or classical studies. Whatever the practical evidence that was offered to schoolchildren in the form of experiments, documents, maps, costumes, photographs or archaeological remains, it was always used to illustrate a concept or a theory which was also explicitly presented. In other words, it was evidence in support of taught theory and not designed as a child-orientated exercise in research. Until quite recently experiments performed in science lessons almost always fulfilled this supporting role; the theory was announced and then practical illustrations were paraded in its honour – some of them being notoriously obstinate in the yielding of obeisance! However, even if the experiments were not successful, it was a matter of no great importance, since all the details of procedure and the 'correct' results were to be found in the textbooks, which were required reading for all pupils. The discovery or establishment of a new theory was clearly far removed in spirit from any of the experiments that were taking place in the school laboratory.

## Armstrong and the Dawn of Discovery

The first attack on this approach to science teaching was made by H.E. Armstrong at the turn of the century. During this period, when Baden-Powell was trying to stimulate initiative and self-reliance in the young by his invention of Scouting for boys, Armstrong was attempting to instil the same spirit into the conduct of school experiments. It was an appeal for a new attitude towards education which sounds as fresh today as on the day it was written:

> Let it be realised that an experiment is something altogether different from a demonstration or verification, just as a trial is very different from an execution. . .The one involves prolonged mental activity, the other mere mechanical obedience. In schools generally the work done is scarcely ever proper experimental work but merely work involving practical demonstrations or verifications – executions not trials. Much nonsense is talked by trainers of teachers and by not a few teachers who ought to know better about the impossibility of children doing 'original work'; it is forgotten that every conscious act done in ignorance of its consequences but with a distinct object of ascertaining what will happen is an act involving original inquiry.[3]

Suddenly it became possible to see a new dimension to education where experiment could be re-invested with all its original excitement and uncertainty. Not only would it be a real inquiry in itself, but it would also lead directly towards an understanding of theory. At one stroke this might both abolish the boredom of didactic instruction and also introduce the child to the new role of a scientific discovery; scientific practice elucidating scientific theory. Admittedly it was a very slow method of covering any syllabus, but for a while it did seem to be gathering support.

In the wider context Armstrong was no lone pioneering figure; it is possible to see him representing within science teaching an amalgam of many contemporary attitudes, those concerned with general child education, with the history and philosophy of science and even with the current psychology of learning. His outstanding importance, of course, lay in his application of these trends to the teaching of science.

## The Naturalist Movement in Education

Froebel, the originator of kindergarten education, and his mentor Pestalozzi had worked and written about their new liberal methods of teaching early in the nineteenth century, but it was only in Armstrong's time that their influence came to be widely felt. There were many strands of Froebel's system, but the one on which the 'naturalists' of the twentieth century laid most stress was the importance of allowing the latent powers of a child enough room to develop. It seemed to them that what children most needed was simply the freedom to cultivate their own individuality and exercise their own capacity for discovery. Some individuals, like Tolstoy, went so far along this path that it became hard to see much point in the continued existence of the teacher at all. Armstrong himself had been heard to complain that over-didactic teachers were one of his chief stumbling blocks but, as we shall see, he did have a definite task for instruction in his system. However, it is easy to understand how a liberal, discovery attitude towards education might well end by making the teacher seem an outmoded and redundant commodity. In more recent times also the doctrine of free discovery has often been linked with a movement to belittle or even to sweep away the interfering figure of the teacher.

## The Philosophy of Common Sense

By the first decade of the twentieth century the history of science seemed to present a gloriously uninterrupted march of progress. In particular, John Dalton's atomic theory and Charles Darwin's theory of evolution presented enthusiastic philosophers of science with a model of excellence for the discovery of truth in the natural world which they were anxious to transfer to every other field of study. The influence of optimistic scientism spread widely, from Bible criticism to social welfare, and a sadly mixed blessing it proved to be. Herbert

Spencer, who wrote copiously about education, was also the author of an obnoxious brand of Darwinism in which he attempted to justify the absence of welfare for the urban poor by using the 'law' of the survival of the fittest.

The scientific method by which these prestigious laws had been discovered seemed to be no more than a simple mixture of observation, measurement and reasoning. T.H. Huxley, another colleague of Darwin, expounded a view of science which much influenced Armstrong's thinking. 'Science', he wrote, 'is nothing but trained and organised common sense.' The philosopher Karl Pearson had written in 1892, 'We must carefully guard ourselves against supposing that the scientific frame of mind is a peculiarity of the professional scientist.' If this were so, there was indeed no need for specialised teaching, only the encouragement of organisation and practical training as Armstrong had proposed.

It also followed that the scientific method was similar to the best methods to be used in history, politics or the law. Skills gained in one subject could be directly transferred to another. The net result of education, they argued, was a trained mind – 'education is what is left when all that has been learnt in school has been forgotten.' It is easy to see how welcome such attitudes would be to the traditional classics teachers in the public schools. Why then teach science at all? This was not a paradox that worried Armstrong himself, but many of his supporters struggled manfully with it on behalf of the new school study of science.

## Training the Faculties

In the nineteenth century the infant science of psychology sought to divide the domain of the intellect into *faculties* which could be treated as separate and distinct elements. The task of education was to develop these faculties – the will, the imagination, the power of reasoning and the memory – by a carefully balanced programme of activities which was often compared to the training of an athlete's muscles.

Armstrong and other reforming educationalists of the time were expected to justify their new courses by reference to these faculties. In 1896 the Oxford and Cambridge Examination Board gave their approval to a school science syllabus which had very largely been designed by Armstrong. Two years later he published an account of his method under the title *Heuristic Method of Teaching or the Art of*

*Making Children Discover Things for Themselves.* The use of the word 'making' is slightly forbidding; how it must have jarred upon the ears of contemporary followers of Froebel! Armstrong firmly believed that children could be *trained* to discover, and that the process involved the faculties of observation, reasoning and memory. His pupils were set a programme of practical exercises to perform and Armstrong argued enthusiastically for the value of experimental work as a training ground for the mind.

It was fundamental to Armstrong's whole position to believe that there was indeed a heuristic method in which pupils could be instructed. Few nowadays would be so sure that scientific discovery always chugs forward on such definite tramlines; in Armstrong's day this belief rested upon a confident and simple philosophy of science. The subtle and revolutionary changes of scientific theory in the present century were yet to make an impact and there seemed to be no mystery about scientific discovery. Urging children to form conclusions from the results of their experiments thus appeared to be both an ideal educational process and also a likeness to science itself.

Highly trained faculties were regarded as the pinnacle of the educational achievement, producing an intellectual elite who were ready and able to transfer these qualities of mind to any other understanding. So ran the credo of Victorian education – and the Foreign Office, the Indian Civil Service and the armed forces floundered or flourished in its wake. The send-up by W. S. Gilbert of the 'Modern Major General' speaks volumes for the depths of inappropriate learning such a system could produce!

## Changing Times

This early period of educational reform is very instructive. We are now at a sufficient remove from it to observe the curious, inevitable way in which it included and complemented so many facets of current thinking. The moral speaks for itself. However isolated and intimate is the laboratory situation in which pupils and teachers meet during science lessons, they can never be completely insulated from outside influence. Teachers are notoriously impatient of the 'highbrow' theories of philosophy and psychology; this may even be a good and necessary defence of their own intuitive powers of reading their pupils' reactions. Nevertheless they too, as creatures of their age, are bound to be coloured in their thinking and their teaching by current attitudes towards both scientific knowledge and appropriate learning methods. Throughout this book there will be references to such contemporary viewpoints, apparently well beyond the domain of the school laboratory and certainly not obviously important for the planning of next week's lesson on human reproduction or the transmission of sound. This needs no apology. Appreciating the power of this continuing interaction lends a sturdiness to the teacher's basis from which he can innovate with greater confidence, and may add an extra dimension to the science which his pupils learn.

Then Armstrong's method began to fall from favour; by 1925 even he was ready to admit that it was out of fashion. There were probably a multitude of reasons for this and some of them are particularly interesting in the light of present arguments about the teaching of science.

In the first place there were complaints from the front line during the First World War that the almost total lack of elementary scientific knowledge was actually endangering the lives of soldiers. War had become more technological and in particular the advent of gas warfare threw up questions of diffusion and of the very nature of gases, about which the majority of soldiers, both officers and other ranks, seemed completely ignorant.

At the end of the war, Higher School Certificate, the forerunner of Advanced Level GCE, was expanded to include examination options in physics, chemistry, botany and zoology. These were intended to be relevant to university work in science and they had a considerable factual content. It soon became all too obvious that Armstrong's prolonged and free-ranging methods provided far too slow a method for covering such syllabuses.

There were also more remote influences from the realms of pure research and philosophy. The 1920s witnessed a shattering revolution in the basic concepts of physics; both quantum theory and relativity turned the older school-learnt theories on their heads and shortly afterwards chemistry and biology were also transmuted by their own quieter revolutions. Even if no immediate changes were made in the school science syllabus (school teaching and their textbooks have a large in-built inertia!) some dim echoes of the noise of revolution were bound to reverberate in the classrooms and laboratories. Science's triumphant march of progress began to look more like the two steps forward and one backwards or sideways of the contemporary tango. It was not the observations or experiments which were at fault, as they might be in the school laboratory, but the interpretations which were to be put upon them. If the findings of famous scientists could be thus reversed, what hope remained that children could make valuable discoveries by their own unaided experiments?

Inevitably this coloured the contemporary philosophy of science. The certain path of rugged common sense no longer yielded certain and unchanging theory, and the philosophy of science began to present a structure of much more subtle hue than the transferable training of Armstrong's hey-day. Nevertheless his initiative never quite lost its appeal, and when the educational system was ripe for another revolution it was ideas very similar to his that rose to the challenge.

## After the War

By the 1950s a whole new battery of influences began to make changes in science education seem almost inevitable. The 1944 Education Act had set up a new system in which the existence of three different kinds of schools was supposed to match the range of potentialities within the secondary school population. The now notorious 11+ examination was relied upon to chop off the academic tip for the grammar schools; the rest were divided between the secondary modern and technical schools. At the same time the school-leaving age was raised to fifteen. To the optimistic it appeared that every child would now be given the opportunity to profit by the education best suited to his ability, always provided that this remained as forecast at age eleven. However socially and academically divisive such a system may now seem, it can be argued that it demonstrated, at that time, a genuine wish to tailor the extended education to fit the children, rather than to force all pupils into the same academic mould. In this sense it reflected a general movement towards wooing the enthusiasm of children rather than coercing their attention.

This was also a period when early specialisation in education was being widely attacked. In public and grammar schools it was quite commonplace for pupils to be forced to opt for the 'science side', or against it, by the age of thirteen or fourteen; the result was an apparently educated population in which many had never studied any science subject after this age. In a famous and influential lecture C. P. Snow attacked the 'Two Cultures' within society – science and the arts – which coexisted without respect for or understanding of each other's intellectual background. When, after a decade of appalling post-war shortages of teachers, equipment and laboratories, the Nuffield

Foundation put generous funds at the disposal of school science in order to inaugurate new courses, the slogan adopted was 'Science for All'. Since these early courses were all designed for the grammar school, which catered for no more than the top 20 to 25 per cent of the school population, this battle-cry may seem more than a little misleading. It was to be understood only in the sense that all those able or lucky enough to be in such a school should have had the chance to savour the delights of school science!

It even proved possible to argue from the continuous upheaval within the world of science that another type of teaching should be adopted. By this time new philosophies of science, such as those due to Karl Popper and T.S. Kuhn, which will be mentioned later, had developed. These systems both accepted as inevitable, even healthy, the confrontations between opposing scientific theories which occasionally led to the downfall of a long-accepted paradigm. Scientific revolutions and somersaults seemed almost more characteristic of science than the cherished Victorian picture of sedate forward progress. If pupils were to be trained for such an uncertain battlefield it could be more important that they were armed with a foretaste of discovery than with an armoury of well-learnt theories which might be superseded tomorrow.

It must be added that there were also those within the new educational movements, perhaps even a majority, who favoured the discovery method for almost opposite reasons. To them discovery was the irreversible method of science and 'to be a scientist for a day', another of the Nuffield catch-phrases, was to find for oneself the way forward by an impeccably reliable route.

In quite another sphere a whole new understanding of childhood learning and thinking about the natural world had been developing. This was the attractive, child-centred study made by Jean Piaget and his followers. By watching, questioning and testing groups of children of different ages he had managed to uncover a vital and unsuspected stage in their development which enabled the unstructured observation of a curious toddler to grow towards the sophisticated formal reasoning processes which are the aim of secondary science education.

Figure 1.1

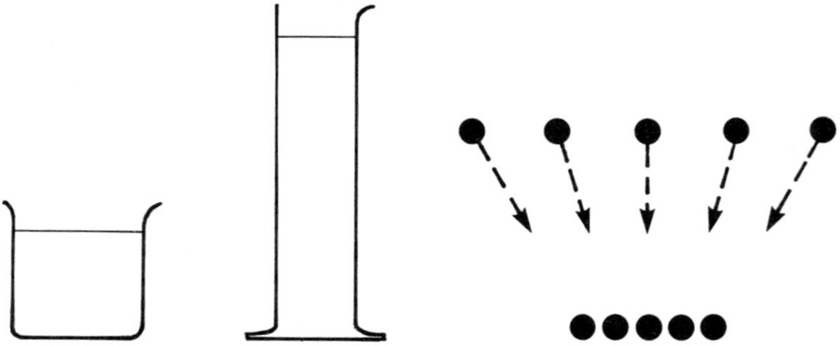

Which holds more water? Which line contains more beads?

Think of *an action* which conects them

In many of the problems and tests that Piaget posed to his sample of children, such as the constancy of a certain volume of water or the number of objects in two different arrays, there always seemed to be an operational stage of understanding when an actual course of action – pouring the water or grouping the objects – became a concrete mental process in its own right. In any explanation it was this action which was visualised and referred to. Purely abstract reasoning, concepts of the conservation of volume or number, arrived later. This seemed incontrovertible proof that the children were right in demanding to play on their own with the intriguing apparatus of the laboratory. In this way they would learn from experiment how to think about natural objects and phenomena before they were presented with the abstract theories of science which rightfully belonged to a later stage in their intellectual development.

## The New Movements in Britain and America

This build-up of pressure came to a head in the late 1950s with the planning of new courses. The next decade proved to be very fertile with a fine crop of new science syllabuses; the most famous of these being PSSC (Physical Science Study Curriculum) and the BSCS (Biological Science Curriculum Study) in America and the Nuffield Science Projects in England. This process is very far from spent even now; every year seems to witness the emergence of yet another important new science course fully furnished with Teacher's Guides, pupils' worksheets and a final examination at the appropriate level. In fact, it has become something of a nightmare for teachers even to keep up with the possibilities available, let alone attempt to assess their merits in the classroom situation. However, since the main theme of this work is a study of various experimental methods rather than a comparison of curricula and content, there is no need to list all of the possibilities now available nor touch upon the vast numbers of 'home-baked' syllabuses contrived by individual schools and offered at O level or CSE (Mode 3).[4]

Parallel to this and intermingled with it ran the other movement, involving a change in the approach to teaching which was to give discovery the pride of place. In the Nuffield scheme this idea was rooted in their much-used word *understanding.* They quoted in its support a supposedly Chinese proverb, 'I hear and I forget, I see and I remember, I do and I understand.' The progress of scientific research, psychological theories of intellectual growth, sympathy for the childhood experience and even the long-neglected teachings of Armstrong had all culminated in a repudiation of the old didactic routine in favour of child-orientated experiment.

A great deal of fundamental work and reorganisation was required. The old criticism of Armstrong's methods still held true, discovery exercises were bound to be slow and need extension over many teaching periods before any useful results could emerge; and these themselves might well prove to be of little weight in terms of knowledge acquired. Their true value would lie in the spirit of inquiry, enthusiasm and individually created understanding that was anticipated. In return for these priceless qualities the new method seemed to demand that the teachers should discard many items of the traditional syllabus; on the other hand, public examinations and the need to meet the requirements of university entrance challenged and menaced any curtailment of the learning process. Fortunately the Nuffield scheme had the resources both to design totally new syllabuses for their O and A level examinations and also the imagination to plan questions which would be in sympathy with the general approach that they were using. After a course which had emphasised understanding based on experimental work, it would have been a travesty of the whole method then to present the fresh, open minds of the pupils with the usual horror of slogging through answers thick with revised and memorised facts. Carefully, and with considerable success, the originators of the Nuffield courses steered their way round the many natural obstacles – and if the result was something of a compromise between free experiment and didactic teaching, at least it was workmanlike and planned to the last item of apparatus to be used and the weekly questions to be set for homework.

Another new feature of these courses was apparent from the first. The Nuffield Physics scheme, for example, was published in 1964 fully furnished with a complete set of Teacher's Guides and pupils' Question Books for each of the five years of the course – and yet there was not one regular textbook. This was deliberate policy. When the Nuffield Project listed the characteristics of the kind of science teaching they wished to promote, they concluded with the following three points:

> a well-grounded understanding of science (or a branch of science), not a knowledge of disconnected facts;
> encouragement of children to think freely and courageously about science in the way practising scientists do;
> experimental and practical inquiry for children as a means of awakening original thought.

This understanding of science was to be achieved in the first place not by reading about theories but by performing experiments and creating concepts at first hand in the laboratory. In this way the children would make their own discoveries and be sure.

In America at this time there was not so much an educational revolution as a public convulsion at the state of science teaching which had been sparked off by that colossal affront to national pride – the Russian launching of Sputnik. School and college science teaching were quickly singled out to be the public scapegoats for this humiliation. Where the Nuffield Project was staffed largely by school science teachers seconded for the purpose from their usual work, in America their colleagues were castigated and then ignored by the new movement which was generated by college educationalists and working scientists. Both the newly created National Science Foundation and the National Academy of Science launched urgent projects for curriculum reform and in 1961 a most forceful and prophetic lecture by Joseph J. Schwab, Professor of Education at Chicago University, heralded a movement for 'Inquiry Methods' in science teaching. The stern admonishment of teachers in the name of national welfare is clear from the opening sentences:

> Teachers and educators concerned with science face a new situation. They are being asked to fulfill an urgent national need, to act as executors of a public policy which is not of their making. They can no longer treat their duties as determined only by themselves.[5]

Schwab's analysis was based on a delineation between 'stable' and 'fluid' inquiry, similar to T.S. Kuhn's descriptions of 'normal' and 'revolutionary' science. When science is going through a normal or stable phase, the current paradigm of the subject, as developed in the contemporary textbooks, dominates all experiment and theory. More detail is added to the paradigm, but the whole activity is only to be reached through an intensive formal education in its operation. Fluid or revolutionary science, by contrast, only takes place at those moments in history when the previous paradigm is discredited and a new one is set up in its place. Examples include the Copernican revolution in astronomy or the wave mechanics revolution in electron theory. For such sparkling innovations it is not so much formal training in an established paradigm that is required as the creativity of new imagination – a far more difficult objective for any educational programme.

Schwab maintained that the rapid growth of science had so shortened the time span of stable inquiry that it was no longer viable itself as a guide for science teaching nor valuable as a source of technological innovation. Fluid inquiry, on the other hand, showed scientific knowledge as uncertain and hard to come by. Many of Schwab's examples were aimed at college students via the study of original papers. In practical work he aimed to give first-hand experience of the difficulties of specifying an actual problem and of collecting data. He too agreed that experiment should precede classroom instruction, and that practical manuals of instruction should be replaced by 'permissive and open materials which point to areas in which problems can be found'. He also discharged a volley of disapproval at the 'rhetoric of conclusions' contained in the usual student textbooks.

A comparison between this seminal lecture and one delivered on behalf of the Nuffield Teaching Project in England, say the well-known paper by Professor E.M. Rogers first read to an ASE meeting in Birmingham in 1964, demonstrates vividly the effects of this divorce in America between the practice and theory of teaching. Schwab concludes his lecture:

> treatment of science as enquiry consists of a treatment of scientific

> knowledge in terms of its origins in the united activities of the human mind and hand which produce it; it is a means for clarifying and illuminating scientific knowledge.[6]

Compare that with the exuberant and child-centred conclusion to Rogers' paper:

> in teaching physics for understanding, we hope for great results in this generation of children but greater still in two generations, when these children can tell their children in turn 'Science is delightful, interesting, powerful; science is great thinking and clever doing; science makes sense.'[7]

Despite this contrast the upshot of these two movements was similar. School textbooks were unanimously cast as the villains of the piece and were rejected as a respectable teaching implement. Experiment was promoted in their stead to pride of place in the teaching armoury either as inquiry into a problem posed by the teacher or, even more boldly, as an open-ended encounter between pupils and materials which was to generate problems, concepts and results. The role of the teacher was less clear. In England he was given the task of radiating enthusiasm and encouragement, and was even allowed to adapt the course as he went along. In America he was to rally to the national call, to read widely in the philosophy and history of science, and to carry out the behests of the educationalists.

## Time for Review

The lifetime of Armstrong's early initiative in education was about twenty years, and by now roughly the same span has elapsed since the 'golden years' of Nuffield. Once more, as time slips by there are subtle changes in our educational aspirations, our philosophy of science and our attitudes towards teaching which should enable us to take up a stance which is both critical and constructive.

We could begin by looking at those features of this attractive discovery method of teaching which have cried out for examination from Armstrong's day to the present. Admittedly it is an emotive area where thinly disguised terms of specialised abuse are apt to be hurled about in a blindly ideological way. 'Authoritarian', 'traditional', or 'radical', 'undisciplined', they cry! Just because the teaching of children is so important and so close to the heart, hardly to be attempted without care, sympathy and personal involvement, there is no reason why such awkward questions should not be asked and answered as far as is possible.

Are there any heuristics of discovery? Can one actually train children to discover? Are they to be guided to the brink or should they be left free to choose both the problems and the methods that they want? Are there different degrees of freedom within what we call discovery learning?

Where does the teacher fit into this? Would the children have greater space to exercise their valuable initiative without his presence? Is a worksheet of written clues better than an authoritative, talking teacher?

Should children be encouraged to work at their own individual pace? What place, if any, does the social interaction between children have in the discovery sequence? What can they teach each other?

Must the syllabuses be jettisoned? Is it possible to contrive that a sequence of discoveries will cover the appropriate topics? Can new concepts and theories also arise from practical experiments without exposition from a book or a teacher?

Does practical laboratory work itself need instruction like carpentry and metal work? Are there skills to be learnt by detailed instruction and demonstration, or will free experiment and practice prove to be its own tutor in such matters?

What about the hoary question of transferability? Is there any evidence that the initiative encouraged in one lesson transfers to another topic or another subject? Does an act of personal discovery make learning and remembering easier?

All these questions will be touched upon in what follows. Sometimes there will be educational research to cast light upon them or new theories of learning by which they can be studied. In many cases, however, there is only grass-roots experience in the teaching laboratory to guide us; but this is not a field to be despised; this is the arena of the children themselves.

Any scheme of science teaching has two different sets of critics who stand poles apart and only too often make hardly any contact. At one extreme stand the professional lecturers on education; at the other stand the children, uniquely cast both as guinea-pigs and continual assessors of the courses. Between these are to be found the harassed, practising schoolteachers whose opinions are swayed now by one side and now by the other. In as much as they take an up-to-date, professional view of their task they will have imbibed some of the new philosophies of science education by spare-time reading and 'in-service' training, but their daily contact with pupils cannot fail to have seasoned their study with incidental surprises. Because teaching is not an intellectual pursuit, either in the field of education or of science, but rather a continual and intimate struggle to understand, stimulate and assuage the curiosity of the young, these teachers have the opportunity to interpret the interaction between children and science in the battlefield of the laboratory. Released from the dreary necessity for producing an atmosphere of chilling silence while notes and sample problems are taken down at dictation speed, the teacher can now permit, or even encourage, a chatty background to the group experimental work. At once a wealth of unselfconscious comment becomes audible. Mixed though it may be with local gossip and pop-culture crazes, this is immensely valuable source material. Argument, discussion and appeals to the teacher for arbitration show very clearly the efficiency of the methods being used and any difficulties that are arising. Boredom is unmistakable and both absorption and enthusiasm are contagious. If one sort of experimental approach succeeds more than another, then neither the teacher nor the pupils can be unaware of it. Children clamour characteristically for more of what they have enjoyed and criticise with uninhibited frankness what they have disliked or found incomprehensible.

With such ready critics to hand there is no need to stop at an examination of the discovery method. There are other areas of practical work that require investigation every bit as much – demonstration experiments by teachers, experience exercises and exhibitions of specimens and materials. Almost all the published courses include such activities, although not all the Teacher's Guides are as clear as they might be about the purposes behind these different kinds of experiment. Nor is it enough to consider the pedagogic purpose alone; the impact of such lessons in terms of class reaction and behaviour will be an essential, even devastating, aspect of any actual lesson.

In a wider sense the childhood experience of learning science will require some kind of programme not only to provide a steady crescendo of understanding which is effective and eloquent in its results, but also one which matches the teacher's own philosophy of science. This is inevitable; one can only sell a product effectively if one has a strong belief in its merit. For this purpose it will be illuminating to touch upon the experience of research at the front line of knowledge and upon the current theories of the philosophy of science. Once again, as these are applied within the classroom, the comments of our young criticscum-guinea-pigs can be relied upon to give the thumbs up, or down, on any new method which is attempted.

The elusive but central motif of this discussion will be the power and training of the imagination. Few scientists would doubt the importance of this quality in understanding and interpreting the models of science all the way from animal transport systems to the kinetic theory of gases and black holes. Still less doubt exists about the vividness of children's own powers of imagination. These are demonstrated in everything they do from playground games to painting and story-telling. Yet, within science, this wayward faculty needs to be fined and sharpened to a purpose by way of concepts which need definition as well as the visualisation of mechanisms.

How then does the solid hardware of the laboratory, the prosaic bunsen burner, flask and ammeter, relate to a developing imagination? If we ignore this apparently ill-matched liaison between apparatus, measurement and imagination we shall reduce experiment to a level of importance on a par with cooking cakes to a given recipe.

## Notes

1. Isaac Todhunter, *Conflict of Studies and Other Essays* (Macmillan, London, 1873). Quoted in Robert L. Weber and E. Mendoza (eds.), *A Random Walk in Science* (Institute of Physics, Bristol, 1973).

2. It is probably unnecessary to point out that such schools as did offer science were largely the well-established public schools. Indeed the forerunner of the present Association for Science Education was named 'The Association of Public School Science Masters', which had a predictably elitist view of its objectives. The first president of this association is on record for stating the aim of school science as 'the means of sifting out from the great mass of the people those golden grains of genius which are now too often lost among the sands of mediocrity'.

3. H.E. Armstrong, Address to the Portsmouth Secondary Educational League, 1908. Quoted in G. Van Pragh (ed.), *H.E. Armstrong and Science Education* (John Murray, London, 1973).

4. The Certificate of Secondary Education, which was introduced in the late 1950s, was designed to be an examination for the 20-60 per cent range of ability, but was intended to be more strongly based on school work than the Ordinary level GCE. In Mode 1 examinations the syllabus is set and examined by the Regional Board, but includes an assessment on school work. In Mode 2 it is the school that sets out the syllabus, although the papers are still set and marked by the Board. In Mode 3 the syllabus, the setting of the papers and the marking of them are all carried out by the schools and merely moderated by an examiner from the regional Board.

5. Joseph J. Schwab, *The Teaching of Science* (Harvard University Press, Cambridge, Mass., 1962).

6. Ibid.

7. Professor E.M. Rogers. Paper given at an ASE Meeting, Birmingham, 1964. Quoted in *Nuffield Physics Teacher's Guides* (Longman/Penguin, Harlow/Harmondsworth, 1966).

# 2 DISCOVERY AND PLAY

## The Ideal

Children at play do make discoveries. Maybe it does not happen very often and perhaps the inarticulate mind of the young child loses the wonder of the moment all too soon in another busy activity; but to those of us who have been privileged to watch the solemn concentration of the game it has a fittingness and a personal triumph that no formal learning process can match. For an example of such simple discovery one can hardly do better than quote from Piaget the childhood experience of a friend:

> he was seated on the ground in his garden and he was counting pebbles. Now to count these pebbles he put them in a row and he counted them one, two, three up to ten. Then he finished counting them and started to count in the other direction. He began at the end and once again found he had ten. He found this marvellous. . . So he put them in a circle and counted them that way and found ten once again.[1]

This sounds like the prototype of all experiment and the astonishing conclusion discovered by this four-year-old boy is indeed a vital landmark in understanding that all children must achieve. Certainly no ordinary parent or teacher sets out to instruct a young child that the number of objects in a set always remains the same. It is considered enough merely to recite the list of numbering words to a listening child, accompanied by a ritual of touching coloured bricks, beads or fingers, and then encourage him to repeat the sounds. The truly enormous task of understanding number is usually left for the child himself to unravel by private play. There is now ample evidence that this free activity is a vital factor in the mental development of all children. Those deprived of the opportunity for play inevitably suffer some intellectual retardation. Piaget called this 'logical-mathematical experience' and maintained that, for the understanding of abstract concepts like the permanence of number, it was of far greater value and effect than passive perception – looking at the colourful number patterns illustrated in children's books.

It is easy to see the attraction that this kind of unprompted discovery has for all reformers of the teaching of science. It is self-motivated and natural, doing away with the artificiality of command and coercion. It generates real absorption and concentration and its successful conclusion brings a delighted understanding which is personal and immensely satisfying. Best of all, it seems to teach science by science's own particular method – experimental discovery. Even if the whole of school science learning cannot be achieved by a series of regular flashes from such moments of illumination, surely we should try to incorporate some of this kind of activity into science lessons? The attempt poses many problems – the number and disparity of the pupils, individual choice, lack of spontaneity, and, most urgently of all, the role of the teacher himself.

## Away with Schools and Teachers

In the past the teacher has always had a recognisable brief, to instruct children along the lines of a fixed syllabus of study. For those who have unhappy memories of school it is all too easy to see the teacher as a tyrant crushing all initiative and curiosity out of his terrorised charges in an attempt to force them through an incomprehensible and unrelated course of study. It was revulsion from this picture that brought into being the progressive movement in education which has argued, since the days of Dewey and before, for child-selected topics and freely chosen projects. Its advocates maintained that no final reward of high income or social status could ever compensate, even the successful survivors of the traditional grind, for the ignominy of bullying or for the loss of delight in learning. Certainly if the child is to achieve happy and spontaneous discovery it will not be by instruction. For some of the extremists who cherish this special goal it is inconceivable that any school or any teacher would permit the flowering of such a slow, idiosyncratic process. The results of this pessimism are the 'free schools' of England and the 'de-schooling' movement in America. Nevertheless, despite the passion and polemic, no real alternative to the school has been proposed except the group or mini-school, and an alternative to the teacher has proved just as illusory. No one has seriously championed a return to solitary learning by 'sitting by Nellie'. All are agreed only that the new teacher requires a different approach to both study and children.

> Education defined as planned learning implies. . .a two-person relationship between 'teacher' and student. In terms of the commonly held theories of learning, the role of the 'teacher' is to select the learning task, to select the appropriate reinforcement schedule, and to supervise the learning process. Since, however, the aim of education is further learning the best education is that which maximizes lifetime learning. A program for lifetime learning can be written only piecemeal, and it appears, intuitively, that the later stages of an optimum program can only be written by the learner himself. One objective of education, therefore, is that the student become his own teacher.[2]

The argument made here about the continual process of education throughout life is now becoming very familiar. For a country in the throes of a post-industrial revolution when automation can be expected to remove the necessity for much unskilled labour and, possibly, increase the number of jobs available to those with appropriate information, the point is important. Whenever this topic raises its head in the education world, it is always in conjunction with a change or demotion in the role of the teacher. Very rapidly this subject retreats out of the area of child discovery into that of adult retraining, and even if the connecting links are discernible, it is better treated in a later section (Chapter 4).

The second point of the argument concerning the 'piecemeal' nature of the learning programme can be more worrying. The topics which present themselves to the learner may well be a chaotically random collection. Discovery, as we have seen, has this element of chance and serendipity, and when the full understanding of an experimental situation depends upon a whole series of previously learnt concepts, such an unstructured learning programme could well be totally unhelpful. There can be no doubt here; a pure discovery programme does away with every vestige of planned syllabus. Its attractions and benefits must be sought in quite another domain.

## Through Children's Eyes

Teachers and educationalists often wax almost lyrical about discovery; they may be remembering a golden moment from their own youth or romanticising an incident in the history of science. Yet whenever we find an instance of genuine discovery, untrammelled by hidden organisation by a teacher, it has qualities of childishness, in its true sense, which appear to be at a very far remove from the teaching laboratory.

Teaching, like raising children, is an art which requires moments of *dégagement* as well as periods of instruction, a capacity for watching without intruding into a different world of feeling. Here are to be found the sudden promptings of childish imagination, possessiveness, rivalry, invention and wonder. Only if they are accorded some respect, or, at a very minimum, are not ridiculed, will the child retain enough confidence to follow his own path towards occasional personal discovery.

The following verbatim account of a familiar, almost hackneyed, experiment is included to illustrate the intrusion of childish feeling into school science:

> when it had its minute of boiling it steamed out happily then suddenly as the heat was removed it had a cork put on it, it soon found that as it cooled it didn't have enough energy to let off steam. When it was full of expanded hot water, full of steam it changed from this to the expanded water subsiding, the steam getting very deflated not warm any more and water drops rolling down the insides of the walls of the can. So the little empty syrup tin would have had a huge vacuum inside had not the atmosphere caused it to heave its sides in and go 'gnab' [this is a Nuffield joke about the opposite to a bang].

Here is a dramatic experiment from which it is often argued that children will 'discover' the force of atmospheric pressure – not a bit of it! Instead the pupil, who after all neither dreamed up the question nor performed the experiment, has retreated into delightfully animistic feelings which are far more concerned with the tin's unhappy deprivations than with the unseen air pressure on its external surface.

When we attempt to see the emotive force of discovery through children's eyes, its main feature is one of pure *surprise* and astonishment. If an adult researcher makes a fine experimental discovery it is not in an intellectual vacuum; he is working within a framework of theory, hypothesis or inspired guesswork which enables him to anticipate the conclusions that he may obtain. He then recognises his success with Archimedes' cry of 'I have found it. . .(the effect I was looking for).' A child, on the other hand, greets his discovery with 'Look! This is extraordinary, wonderful!' Preparedness is essential for appreciating an experimental breakthrough in real science without which it has no significance,[3] but the young child is untrained in both the formulation of scientific hypothesis and the rationale of experimental testing. He looks only for something new and sensational.

**Surprise or Confirmation?**

When the discovery method was first advocated by Armstrong it was indeed a simple training and observation process which could easily become a teaching programme. In the 1950s, when it re-emerged, it was seen as a far more complex activity, even including guesswork, and yet there was still this tacit assumption that it could be taught to children. As the following passage shows, it was held to be an attitude rather than a discipline.

> Mastery of the fundamental ideas of a field involves not only the grasping of general principles, but also the development of an attitude toward learning and inquiry, toward guessing and hunches, toward the possibility of solving problems on one's own. Just as a physicist has certain attitudes about the ultimate orderliness of nature and a conviction that order can be discovered, so a young physics student needs some working version of these attitudes if he is to organise his learning in such a way as to make what he learns usable and meaningful in his thinking. To instill such attitudes by teaching requires something more than the mere presentation of fundamental ideas. Just what it takes to bring off such teaching is something on which a great deal of research is needed, but it would seem that an important ingredient is a sense of discovery – discovery of regularities of previously unrecognized relations and similarities between ideas, with a resulting sense of self-confidence in one's abilities.[4]

This quotation which is from a short but very influential book written in 1960 by the American educationalist Jerome Bruner shows clearly that the model of discovery he had in mind was simply a junior version of adult research. He assumes that a child could and would orientate his inquiry towards finding these regularities in nature. A sympathetic following of children's activities shows that almost the exact opposite to this is actually the case. What children most hope to discover is the irregularity, the unpredictable, the surprise.

There is also a circular element in this argument which was so much used to promote discovery teaching. If, as Bruner argues, a science student needs to understand something of the discovery attitude in order to make science meaningful to him, it seems odd that he is expected to be able to use it in order to discover it! Teachers are not expected simply to instruct their pupils in the operation of discovery, it is to be a subtly induced change of attitude. Indeed the method of science is itself to be discovered by the pupils' undirected discovery.

Bruner's emphasis on the element of searching for *expected* regularities in scientific research is true at two different levels. In the widest sense the scientific endeavour would not exist at all if this expectation lost power, but in a narrower sense it leads to the very antithesis of a child's idea of discovery. In 'normal science', according to Kuhn's classification, when those involved in research are totally immersed in the accepted paradigm, the general outline of the expected result of an experiment can be so clearly held in mind that there is no room left to entertain the possibility of something new coming about.

This can be illustrated by a curious anecdote from early nuclear research. In the 1930s there was a rush to perform the new 'atom-bashing' experiments with the recently discovered neutron. In most cases this particle was momentarily absorbed into the nucleus of an atom giving an unstable isotope which then decayed radioactively, emitting one or more low-energy particles which could be detected with a Geiger tube. In Cambridge it so happened that a young research worker set up a target composed of uranium atoms and found that the needle of his instrument occasionally flicked wildly off-scale, indicating emission energies far in excess of anything that current nuclear theory could predict. This result was not within his area of preparedness and, though certainly a surprise, it was received merely as an irritating practical hitch. He assumed that his instrument was misbehaving and set about 'fixing' it so that such unbelievable readings could not occur again. In this way he managed to avoid discovering nuclear fission!

The moral of this story might be that the element of surprise is as vital to progress in adult science as is child science, or it might simply be an object lesson in the necessity for careful control, experimental repetition and the preservation of an open mind. Hindsight criticism of research is not the object of recounting this tale, which can be matched many times in the history of science. It is used only to illustrate the fact that total surprise is suspect and even unwelcome in a real research situation, whereas to a child it is the delightful and sensational prospect of every new experiment. This, no doubt, is a major reason why chemistry and home-made explosions are so popular among early teenage children!

It would do less than justice to the sensitive writings of Jerome Bruner to give the impression that he totally ignored this element of discovery. In an early essay on creativity he wrote, 'An act that produces *effective surprise* – this I shall take as the hallmark of a creative enterprise.'[5] He then explained that he was using the word in a special sense as that which 'produces a shock of recognition following which there is no longer astonishment'.

This is likely to be the kernel of the kind of discovery that teachers most wish for their pupils. The surprise is not to be just a delightful but dead-end experience; it has purpose, is effective and results in the understanding that such recognition brings. When the brief moment of surprise is over the pupil's educated response is, 'Oh yes, of course, *that's* how it all fits together!' This harks back to Bruner's earlier point about searching for orderliness and regularities in nature. No one would cavil with this – it is one of the very highest objectives of all teaching and the moment when a pupil first realises that there are systems of thought which make connections between different scientific phenomena is very sweet indeed to his teacher. However, it is not usually the result of *undirected* experiment.

Bruner also wrote that 'discovery, like surprise, favours the well prepared mind'.[6] This seems a curious point of similarity: discovery, as Columbus found, requires a great deal of preparation, but surprise is the gratuitous reward of ignorance. Similarly in science lessons a well-designed, deliberate class experiment which builds upon previous knowledge can push the prepared minds of children into finding out that the burning of foods produces similar products to respiration, or that the colours of the spectrum can be recombined to give white light. This sort of activity is in quite another class of 'discovery' from the free play of children and rightfully belongs to our next section. Here we are concerned only with child-generated discovery.

There were four main benefits which Bruner claimed would accrue from using a discovery method of teaching:

(1) increased intellectual potency;
(2) a shift from extrinsic to intrinsic reward;
(3) learning the heuristics of discovery;
(4) an aid to conserving memory.

After more than a decade of practice considerable doubt has been cast upon these claims. We shall see in the next chapter that, even when the discovery is of a highly teacher-orientated type, educational research has not found intellectual potency (as shown by the ability to solve problems), nor has the power of memorisation shown the expected improvement. Of course it is true that children can be instructed in a particular method of procedure, the 'discovery method' or any other. This begs two questions: is there any such thing in scientific research? And did Bruner intend that the children should be *instructed* in the method that they were to use? On the other hand, true, open-ended childish discovery is of its nature haphazard; it can have no heuristics except spontaneous feeling.

Only the second of Bruner's objectives survives quite untarnished. By its very nature this kind of inquiry would have to be self-motivating; it alone of all possible laboratory activities aims at no pre-set goal and wins no reward in the shape of high marks or approval. Teacher praise or comment designed to reinforce a familiar or promising line of inquiry, or to discourage an apparently pointless or repetitive activity, is not only irrelevant, it devalues the whole tender effort at independence. It is no mean achievement for the busy organised pupil to find time and space to enjoy his own endeavour without reference to some external standard of what is right or useful. Essentially it is *play*. As Koestler has written, the basic characteristic of such 'ludic learning. . .is the self-arousing and self-rewarding nature of the activity'.[7]

## Of Chance and Play

Most teachers have observed the start of such spontaneous play, refer to it as 'messing about' and, under the pressure of a pre-structured lesson full of necessary objectives to be achieved within a limited time, tend to discourage it. This begins after some organised practical exercise has introduced the pupils to the use of some new apparatus. Then, having acquired the necessary confidence in its use, a sense of fun, inventiveness or raw competition can turn the sober experimental procedure in a totally new direction. A prism may be rotated in the path of a ray of light until the two halves of the spectrum emerge from opposite sides. An unfocused microscope can merge a blurred image into the outline of a distant window. A plastic spoon is watched with glee as it slowly droops and melts near a Bunsen flame. Any specially good (spectacular) results are loudly and triumphantly demonstrated to neighbouring groups and under the spur of 'game-making' competition may enter the activity. An empty flask with a long tube which has been used to demonstrate the expansion of heated air into a beaker of water now by chance includes a small water index and becomes the basis of a race to find out whose hands are hottest. If they join two such flasks together the game becomes a variation on the traditional 'tug-of-war'. To the watching teacher it is also a differential air thermometer, but no such technical term need be offered. Apart from the pleasure and unforced concentration of the activity, its immediate benefits are only some practical confidence and the kind of free lateral thinking which is essential for all invention.

This strange antic activity arises the more readily because some familiarity with the apparatus has been established, giving the necessary fillip to confidence. In other ways it does not fit easily into the class situation. The varied interests and capabilities of the children become more pronounced as the chaotic discovery games proceed. While some pupils become increasingly engrossed, others will be reaching the end of their invention, and rowdy ebullience or simple boredom may take over. This must seem to be the moment for re-imposing class teaching for the sake of the idle, even though others still hanker for more time to preserve the impetus of their own discovery – conflicting needs which demand different solutions. The structured lesson will get under way once more, but the real value of child-generated discovery insists that we do not neglect it in the narrow cause of class discipline. We need to provide generously the necessary time and opportunity for repetition, variation and the possible emergence of further inspiration. If neither home nor library can satisfy this need, then the opportunity for such pupils to work on in the laboratory during lunch-hour or after school becomes imperative. A Science Club with loose supervision and the freedom to use familiar apparatus on one's own is an inestimable gift to those with an obsession for tinkering with practical ideas and one which, on occasion, can inspire real scientific enthusiasm.

There is a most curious relation which exists between discovery and boredom, between enthusiasm and repetition. It is often the lone child with no activity to hand who notices some new potentiality as the result of simple fiddling, and it is the enthusiast who shows an apparently endless desire to repeat exactly the same experiment again and again. Watching them, it becomes clear that the unresolved nature of the wonder is in itself a source of pleasure, like the astonishing sight of a strange animal, a conjuring trick or a colour transformation. The scientific aim to resolve and explain is often less strong. A pupil who stumbled across the effect of electrolysing acetic acid between lead electrodes was happy to watch the formation of the brilliant yellow deposit time and again before delving into books to read up on the subject. Others who became fascinated by the wave-traces of sounds on the oscilloscope and begged for its use almost daily for a month still continued to exhibit and enjoy their old favourites for a long time before embarking on any new patterns. This reaction is probably closely related to other hobby activities such as insect-rearing or stamp-collecting, where the impressive forces of self-motivation can withstand and enjoy a routine which may seem forbiddingly monotonous and repetitive to the onlooker.

## What Use is This in School?

Granted the motivation and absorption, is this unstructured exploration of any real educational value? It has little enough to contribute to the usual learning categories of 'problem-solving', 'concept-formation' or the 'heuristics of research' (if indeed any such exist). It is doubtful whether it can even boast a contribution to that vaguer educational goal of 'cognitive growth'. Sudden spurts in thinking power, like those in bodily growth, seem to have little connection with the quality and quantity of the nourishment taken! Model-making and the following of electronic instructions are the usual activities of a science club and they do allow the children to cultivate a little extra practical know-how. Independent tinkering about with lenses, mirrors or familiar electrical circuits would seem to bring a sparse return. Nevertheless there are plenty of school situations when this pride in 'going it alone', coupled with the ambition to acquire the prestige of a child expert, become the strongest ally a teacher can have in helping the wayward and defiant in his class.

Successful learning requires a very large measure of imitation – although it may be educationally unfashionable to admit this. Recent research has been ponderously examining the effect of children's attitudes to their teachers as a possible motivation for learning and has offered its usual crop of conflicting results.[8] To benefit and learn from a lesson the largest single contributory factor is not intellectual prowess but the simple capacity to empathise with the teacher. This is not the same as teacher-liking. Fortunately most children do have this requirement – they positively enjoy being carried along by the suggestions, directions and the enthusiasm shown, and they easily become a part of the class activity.

Alongside this convenient willingness to conform most children also have the opposite urge – to be differentiated from their peers,[9] to stand out from the crowd, to resist the overt suggestion and substitute their own. These two sets of characteristics are present in all children, but in differing proportions; some are too docilely obedient and lacking in initiative, others are excessively egocentric and resentful of instruction. The former set contains the pupils least likely to profit from the freedom to explore on their own and indeed could be distressed by the very expectation. The latter set are the 'stroppy' pupils who stand out like a sore thumb in every lesson. There is egoism and antagonism here which may well disappear with the passage of a few months, but while the mood lasts it renders all teaching painfully difficult. In these circumstances making up their own experiments is just what their touchy vanity requires. The extra opportunity to contrive and claim a personal discovery is very dear to these children. The skilled teacher will express only tactful surprise and, at most, a restrained and humble interest in the project, but it is a boon to him too. Sometimes this expression of defiance can itself produce the first useful bond of respect between teacher and pupil.

So far only two arguments for the value of this licence to discover have been offered, in contrast to the inflated claims so often made on its behalf:

(1) the free expression of *childish* attributes within the field of school science in order that the young pupils can enter and enjoy the new activities in their own familiar way;
(2) a remedial effect for those children who find the usual imitative posture of the student *vis-à-vis* the teacher too dependent and irksome.

Both of these advantages are specifically child-centred, making no claim to teach the work attitudes of adult scientists. If the aims of science education were simply to train the young to take up their places in the laboratories of the future, then these benefits might appear slim indeed. Seen on a more generous scale of objectives, however, most schools could well tolerate and even encourage some of this peripheral activity in the no man's land between lessons and play without the usual complaint about wastage of precious teaching time.

Is there not a real disadvantage in recognising this naïve activity in the name of science? If these first chaotic attempts at exploration go uncorrected, will it not give a completely false impression of research?

Scientific work is serious and quantified, not exuberant and sensation-seeking. The cautious response to this objection is that later science teaching will soon correct this impression. More audaciously, one could question whether these fresh qualities of childish exploration either are, or indeed should be, totally absent from 'important' research.

**Element of Surprise in Scientific Research**

It is not hard to point out occasions when eminent men of science have behaved with outrageous spontaneity and enthusiasm. Humphrey Davy literally danced round his laboratory giving vent to wild delight at the first appearance of a globule of potassium and was quite beyond the reach of sensible conversation for nearly half an hour, according to his own brother's account. Benjamin Franklin and his friends of the Library Circle of Philadelphia arranged a celebratory 'electrical picnic' on the banks of the Potomac river, eating roast turkey which had been electrocuted and drinking dangerously from the lips of charged Leyden jars! Attitudes of deep solemnity are obviously not essential for the appreciation of science. Many scientists have also been adept at arranging spectacular public demonstrations and, although apparently prompted by a worthy didactic motive, give ample evidence of an almost childlike delight in producing dramatic effects.

Even within the process of scientific work the wide-eyed curiosity of the unlearned is not altogether to be despised. Of course the background to every planned experiment is signposted by theory, hypothesis and counter-theory which are only understood in the light of many years of training. It follows that the possible results are also, to some extent, predicted by these tentative explanations; an event which is expected to occur at a particular temperature may suffer a small but supremely important deviation from this prediction, which would then provide a measure of support for an equally well-known rival theory. All is planned in knowledgeable anticipation. However, it also happens in science that what man proposes God disposes and that there ensues instead the unforeseen, the extraordinary, the surprise. Then the simple ability to watch open-eyed, the capacity to entertain the wonderful and unthought-of are at a premium and sometimes in the midst of this sophisticated over-preparedness they can be very hard to come by.

Consider the early work on low-temperature physics. Kamerlingh Onnes, the first to liquefy the obstinately gaseous helium, spent the three years from 1911 to 1914 examining the electrical resistance of mercury and other metals in the range of temperatures that liquid helium now afforded near the absolute zero. He confidently expected a gradual decrease in resistance and, in spite of a consistent series of careful experiments, he found the abrupt disappearance of all measurable resistance at the onset of superconductivity so inexplicable as to be unbelievable. In the place of delighted triumph at a new discovery there was only worried repetition and a suppression of publication.

There is another example of this inability to see through a fog of learned expectation which is still more remarkable. Ever since Onnes first produced liquid helium, experimenters in many countries had been struggling to reach down to ever lower temperatures by reducing the pressure on the surface of the liquid so that it might cool by the action of its own boiling. In every case at 2.2° above the absolute zero of temperature the familiar ebullition suddenly ceased as the curious liquid continued to boil, but without bubbling! It was an effect so simple and vivid that any watching child would have seen it, as must a host of mature research workers, but its first mention in print occurred only 24 years later. This was more than an isolated, if sensational, phenomenon. It contained a hint writ very large for unravelling the dilemma of heat conduction in liquid helium – but it went unremarked. Every explanation proferred by the formidable mathematics of quantum mechanical liquids was respectfully received, perhaps partly on account of its very difficulty, but the occurrence of an unusual visual effect was suspect, inconvenient and ignored. Oh for the credulous eye of a child!

## Notes

1. J. Piaget, 'Development and Learning' in R. E. Ripple and V. N. Rockcastle (eds.), *Piaget Rediscovered* (Cornell University, Ithaca, New York, 1964).
2. I. Illich, in D. Levine and R. Harivhurst (eds.), *Farewell to Schools???* (Charles A. Jones, Worthington, Ohio, 1971).
3. See Chapter 7.
4. J. Bruner, *The Process of Education* (Harvard University Press, Cambridge, Mass., 1977).
5. J. Bruner, 'On Knowing' in *Essays for the Left Hand* (Harvard University Press, Cambridge, Mass., 1979).
6. Bruner, 'The Act of Discovery' in Bruner, *Essays for the Left Hand.*
7. A. Koestler, *The Act of Creation* (Hutchinson, London, 1969).
8. Kahn and Weiss, for example, 1973: 'whatever else is in doubt, the role of the teacher is crucial in the development of the pupils' affective behaviour over the field of education as a whole'. ('The Teaching of Affective Responses' in R.M. Travers (ed.), *Second Handbook of Research on Teaching* (Rand McNally, Chicago)). Or Omerod, 1975: 'the correlation coefficients obtained for the science subject choice/teacher-liking relationships. . .are negligible in most cases.'

(M.B. Omerod with D. Duckworth, *Pupils' Attitudes to Science* (NFER, Windsor)).

9. Jerome Kagan (Harvard): 'The need for differentiation is important for creativity – to be different from one's peers.' (In Lee S. Shulman (ed.), *Learning by Discovery* (Rand McNally, Chicago, 1966).)

# 3 DISCOVERY TEACHING IN ACTION

## Discovery and the Curriculum

It is not enough to recognise child discovery in science as a kind of glorious messing about in the laboratory on the trail of delighted surprise or obsessive experimentation. 'Discovery' is now a serious teaching strategy which boasts a large following and demands consideration in this context. Since it has already been in action, at least apparently, for well over a decade, not only its aims but also its practice can be observed and assessed. There has been time for some educational research to have been carried out on its learning efficacy and also for a critical viewpoint to develop.

The first and indeed one of the most intractable problems has been how such free discovery can be fitted into a curriculum. If discovery is an essentially spontaneous activity, how can it be accommodated within a planned course of study? It seems an insoluble paradox and to the revolutionary leaders of the 'humanistic' view of education it remains exactly that:

> The humanistic educator must be continually aware of the growing edge of the child's intellect and character. He must be exceedingly resourceful in seizing upon opportunities of the moment to turn them to educational account. . .He cannot work from a prepared curriculum but must invent as he interacts with the child.[1]

There is little in this advice from the 'enfants terribles' of the anti-school movement to shock the modern conventional teacher. The only moment of jibbing comes with the insistence on jettisoning the whole prepared curriculum! Course material is not sacrosanct, the spectre of examinations need not be invoked, but the teacher has another obligation towards his growing charges, to introduce them to the whole range of theories and methods which are their scientific heritage. The task is large and the time is short, but combining these different objectives may not be impossible. If the rewards of child discovery are considered sufficiently valuable to warrant a small contraction in the syllabus, the challenge will become one of providing loopholes within a set course of study where this invention and interaction can arise naturally. Such attempts to encourage children to take the initiative in their own science education have taken place and have yielded practical experience of such 'discovery' teaching in action.

The best attempts which have been made so far to open out these free spaces within a planned science curriculum occur in the Nuffield Project's schemes. In spite of a conscious aim to develop the subject in a continuous, logical way, any study will provide moments of hiatus when a new topic is about to be broached. Just because the children have not yet received any instruction on the new material, this might seem an appropriate moment for them to bring to the fore their own uncluttered curiosity and natural inventiveness in a free period of simple discovery. Perhaps this is a cultivation of primitive child science in a way which is similar in feeling to that given to untutored child art. Yet the comparison is not entirely fair, even if it is based on the same sympathy for childhood experience. Child art has no syllabus, few learned skills and little accepted standard of judgement; it results from a continuous growth of play from the earliest years. Child science is different in all these ways and watching a class attempt this form of primitive discovery shows many unsuspected pitfalls.

The earliest of the Nuffield courses was in physics, where the fragmentary nature of the subject at elementary level still involves absorbing the material in discontinuous gulps and, as a breath is drawn, there is an obvious opportunity to slacken the pace of tuition and provide time for free discovery. The first of these natural breaks occurs part of the way through the first year. There has been an introduction to the subject through the beauty of crystal-making, which is used to stimulate interest in measurement and weighing. This is followed by the construction of sensitive micro-balances out of pins and drinking straws. Then follows an 'open experiment' with levers. No statement of the mathematical principle of moments is made; instead the originators of the course spent a considerable amount of time in designing a simple kit of lever and weights (uniform square brass plates) which the children could use in order to uncover for themselves the ancient law of the lever. No explicit problem is stated, no procedure is recommended, the pupils simply play with the apparatus and the law is discovered afresh. That, at least, is the hope.

### Practice and Problems

Practical difficulties of two sorts arise. No lever system balanced below

Figure 3.1

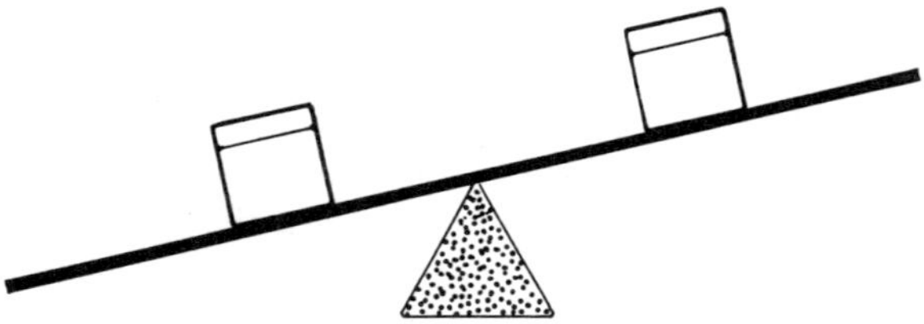

BALANCE?

Figure 3.2

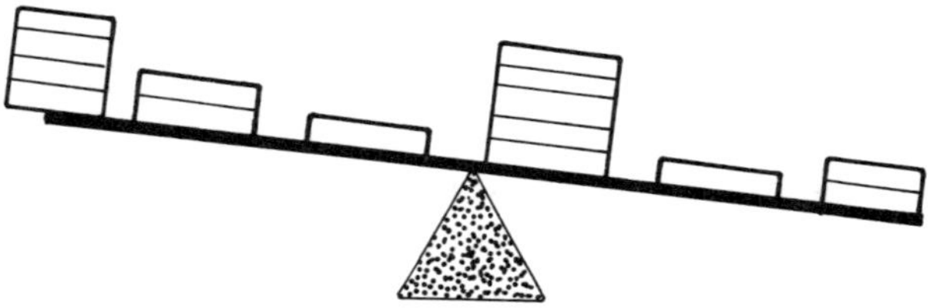

'What does this prove, Sir?'

its centre of gravity can be stable in the horizontal position that all children recognise. A creaking see-saw or a balance with hanging weight-pans performs according to expectation, but an over-balanced lever is out of balance and no amount of reassurance will correct this intuitive and basically correct objection. The planners recognised this failing and supplied teachers with arguments to use and instructions on procedure to give out – with a predictably deadening effect on the pupils' free use of initiative.

Practical experience also demonstrated that the children enjoyed spreading out the piles of shiny metal pieces irregularly along the bar. Balance was achieved, in as far as it could be, but the complexity of the piles now completely obscured the simple result expected. Once more the freedom of the play had to be interrupted to give disciplined instruction if the 'desired result' was to be 'discovered'.

**'Organised' Discovery?**

Honesty forces us to admit that an element of fraud has now crept into the process. If discovery is to have real meaning, it must be a genuinely independent activity neither following instructions nor manipulated from behind the scenes by a conniving teacher. It is, as the de-schoolers so firmly state, only to be brought about by releasing the whole enterprise to the children and abiding by their results. If, as in the previous example, the organisation needs to be tight and the teacher commentary continuous in order to coax a pre-planned result from the children, then discovery becomes an elaborate hoax. This state of affairs is now commonplace because of the respect the concept commands. There are teachers' seminars on discovery, apparatus planned for discovery and textbooks by the yard entitled *Discovering*. . ., *Exploring*. . ., *Inquiring into*. . . and *Finding out*. . ., even though the use of any textbook must be the antithesis of spontaneous inquiry. It is equally true that to hand out sets of apparatus, instruct on how it is to be used and pose a question such as 'What happens to the metal rod when it is heated?' or 'Which metals react with dilute acid?', produces obedience, not discovery.

So far has the real nature of spontaneous inquiry been lost in a general confusion of liberating intentions that it has even proved possible to produce assignment sheets with step-by-step instructions and programmed questions to answer and still to designate them as 'creative' and 'open-ended'. The Scottish Integrated Science Syllabus, which is widely used in comprehensive schools all over the country, sees no contradiction in this approach. In its note on teaching method the opening sentence runs, 'In order to achieve these objectives it is essential that "discovery" methods be used.' Nevertheless the pupils' textbooks illustrate every experiment and summarise all the results obtained, while the worksheets 'have a distinct educational value in that they train pupils to follow instructions'! The planners of this course hoped that class discussion before the sheets are handed out would provide the necessary stimulus for creativity. However, it is clear that the 'free' debate must always be stage-managed, for 'though this discussion must give full rein to the pupils' imagination, and their suggestions must be fully considered, it will eventually lead to methods chosen on the Worksheet.'

The pioneer American teaching scheme PSSC follows a similar format. Again there is both a student's textbook and a student's laboratory guide; once again it is the expressed intention of the originators of the course that, in spite of all the planning and instruction, the children should taste the real flavour of open inquiry, both for its scientific value and as an aid to learning. As in the Scottish Integrated Science course, the reasons given for each experimental set-up are clearly explained and the series of questions on the progress of the experiment are often excellent teaching material which provoke the pupils' interest and participation throughout the practical work. There is a sense in which it may be true to say that the pupil is discovering what will happen in each practical exercise but, in the nature of the method, he can never for a moment believe either that he is on uncharted ground or that the experimental venture is of his own making. It is not my intention to criticise this course, only to draw attention to the curiously diverse ways in which the term discovery is used, often within the very same sentence.

> The laboratory is an integral part of the course. In many instances it is the primary learning source. While the student cannot recapitulate all the discoveries of physics, the laboratory is designed to give the student an opportunity for personal discovery.[2]

Armstrong's famous remark that any experimental work undertaken without foreknowledge of the conclusion is discovery now needs qualification. Modern science teaching has, hopefully, progressed beyond the stage of repeating a known procedure in order to obtain an already known result. Any teacher worth his salt leaves some element of surprise in his experiments. The pupils' accounts are no longer entitled 'To show that. . .' or 'To verify that. . .', but rather 'To find out. . .' Yet, when the whole process is directed and sponsored by the teacher and not the child, it cannot constitute free inquiry in any fundamental sense.

Discovery can be recognised by two criteria. First, the query must spring directly from the children themselves, whether latent in their own private puzzlement or prompted by their immediate activities. Secondly, the way in which they choose to manipulate the apparatus that they use cannot be dictated. It follows that their results will also be unforeseen and should not be evaluated in terms of their approximation to familiar conclusions. To encourage experimental freedom and yet to expect a definite conventional result is the educational equivalent of both having your cake and eating it! The value of discovery lies in the open-ended experience that we try to promote and not in an achievement of the next stage in the curriculum. Hard though it may be for the teacher to forgo his usual controlling role in the learning process, his are not the only problems. The children themselves, curiously enough, are faced by even greater psychological hurdles.

**Unguided and Lost**

These problems can be illustrated by reactions to the next attempt at open experiment in the same inventive Nuffield course that was quoted previously. A little later in the syllabus, but also during the first year of physics, the elasticity of springs is examined and, once again, the introduction is by free experiment without previous instruction. The pupils are simply given springs and asked to find out something about them for themselves. With supreme optimism the Teacher's Guide assumes that each spring will be loaded with weights in the time-honoured way and its extension will be measured – but children are by no means so predictable. A spring is a delightful plaything, it will coil and twist, flick and project pellets. Is there any incentive for examining it in the more prosaic way that the teacher anticipates, apart from the uneasy weight of a sensed expectation? After the initial jubilation, during which some proportion of the springs are irretrievably distorted, the mood of the class changes. 'What are we supposed to do now?' 'What does he (or she) want us to do?' 'How do we get it *right*?' Play may be idle and relaxed at home, but school is known to be for serious learning and only the malicious or insensitive can resist for long the growing feeling of unease. Most children wish both to please and to be successful and, in the absence of either problem or method, begin to feel insecure. Activity comes to a halt and attention wanders. A busy, prolonged involvement in doing your own thing in a laboratory presupposes an intense personal curiosity which has not yet been aroused.

**A Summary of Problems**

It is possible to summarise and learn from the difficulties that have arisen. These can be tabulated under two headings:

(1) teacher problems

*A real cessation of instruction* both explicit and implicit is the only background against which children can follow their own ideas.

*A temporary release from the schedule of the curriculum* may become necessary to accommodate unexpected lines of inquiry.

(2) pupil problems

*Practical difficulties* of every sort arise if unfamiliar apparatus is being used.

*Diffidence* in establishing a personal line of attack in the absence of guidance may be the pitiful result of spoon-feeding or a genuine character trait, but it needs to be recognised.

*Lack of involvement* is inevitable on the part of some of the pupils. It is unrealistic to expect the kind of committed interest that discovery demands from each individual towards the same teacher-selected topic.

**How Pure is Your Discovery?**

Unguided discovery as a format for lessons is not only emotionally demanding on both class and teacher, it also appears in practice and on reflection to be quite foreign to the whole ambience of class-teaching. Where discovery is no longer the free individual activity that was visualised in Chapter 2, it very easily becomes metamorphosed into a more familiar, didactic kind of class experiment. It is easy to see that different degrees of pupil initiative may be included under the general umbrella of discovery, but the suspicion grows that it may be honoured more in theory than in practice. Right at the outset of this whole educational movement in 1962 Schwab himself tabulated the possible 'levels of inquiry' in the teaching of biology (it is obviously equally appropriate to the other sciences) in the following way:

| | Problem | Ways and Means | Answers |
|---|---|---|---|
| Level 1 | Given | Given | Open |
| Level 2 | Given | Open | Open |
| Level 3 | Open | Open | Open |

It was not until 1971 that the actual laboratory exercises of the widely acclaimed BSCS (Biology) and PSSC (Physics) courses used in American secondary schools were examined using this tabulation. It will come as no surprise that *no experiments at all* fell into the 'pure discovery' level 3. Even the more adventurous Nuffield courses could not sustain or use the individualistic, chaotic activities that fall into this category, as we have seen. What did cause a wave of shocked surprise in educational circles was the estimate that over 70 per cent of the practical work in these courses allowed no element of inquiry whatever. It was as though the revolution had never happened. The problem, method and answer were all known to the pupils before the experiment even began! One is forced to assume that the illustrious names of those who originated these courses, including that of Schwab himself, were enough to secure their acceptance at face value for over twelve years of school use, when simple counting could show that only a mere 20 per cent of the prescribed laboratory work achieved even the lowest rung of the inquiry ladder. All this must have been common knowledge to hundreds of working American teachers during this period.

In some way it may have been the very crusading zeal of the authors of these new courses which militated against the practice of discovery. Never before had courses been prepared in such detail and precision; so much apparatus was designed, such detailed teaching notes were produced and pupils' guides distributed that the organisation itself may have dispelled the atmosphere of detached curiosity needed for creative discovery. Such precise instructions for the teacher and the children were themselves to precipitate quite another educational revolution.[3]

A rough count based on the science schemes used on this side of the Atlantic shows the majority of class experiments falling into Schwab's first level of discovery – that highly directed and didactic situation in which both the problem and the method are imposed upon the children with only the final answer concealed as a kind of mild incentive to action. This is the strategy that has been graced with the curious title of 'guided discovery', and which is to be found at work in almost every school in the country. As we shall see, it marries well with the group situation.

The Schwab tabulation proves far too crude a tool for examining and assessing the various science courses. In spite of the almost universal lip-service paid to the discovery ideal, the different courses do have their own 'style' and the *objective* of discovery varies from one course to another. Many of the Nuffield Combined Science experiments, for example, are designed to experience some effect in an illuminating but non-quantitative way so that the answer itself is of less importance than the experience.[4] (For example in measuring 'The force needed to get 1 kg moving' in the near-frictionless conditions of a tray covered with beads or ball-bearings, the actual value measured must be vanishingly small and of trivial significance. It is the *felt* comparison with the weight of 1 kg and the ease of its slow acceleration which is the valuable surprise discovery.) In the Scottish Integrated Science course each experiment is typically more concerned to teach practical techniques and to elicit precise experimental 'facts'. Accordingly the instructions are so worded that the children can be 'guided' into discovering these objectives. Of the newer science courses SCISP (Schools Council Integrated Science Project) may seem to score much lower in the discovery stakes than the other two mentioned, but this is because its explicit aim is to uncover patterns in the natural world and to introduce models which will accommodate them. This is far too sophisticated a task to arise directly from the practical work. Now the object of discovery becomes both the pattern and the validation of a prediction from the model.

**Some Results of Research**

Though few critics of discovery learning have mentioned the practical and emotional difficulties, there have been two other recent sources of objection to it. Educational researchers have carried out measurements to assess the effectiveness of this method, both for 'learning retention' and for enhanced interest in science. Batches of pupils who have learnt the same material by traditional and by discovery methods have been tested in several different environments, but the statistics obtained have often been contradictory and controversial. In 1950, for example, L. Szekely[5] examined the effectiveness of teaching a group of senior pupils about moments of inertia by two different methods. The first group learnt traditionally by reading, learning and discussing the concepts of mass, force, angular velocity and moments of inertia before performing an experiment. They were then instructed to set up a torsion pendulum, as illustrated, and timed its period of oscillation for different positions of the weights, W. The second, 'discovery', group began by being presented with the experiment as an open problem to be investigated. Only later, when these pupils had worked out their own conclusions, were they allowed to consult the textbook in order to obtain more precise explanations of the concepts and problems which they had encountered. Szekely's subsequent tests showed:

(a) the first group scored better in recalling the exact definition of moment of inertia;

Figure 3.3

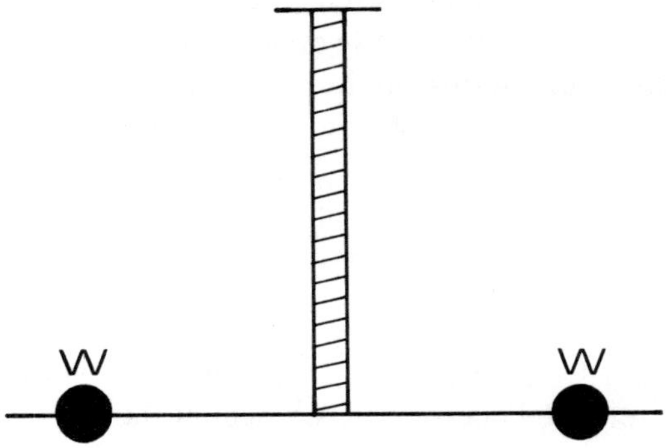


(b) the second group scored better in applying their knowledge to a related practical problem (differentiating between two spheres of equal size and weight, one of which was hollow).

Now these results were precisely those which Szekely and the proponents of the new discovery methods had hoped for and expected. However, two years later the experiment was repeated by a group of three American researchers[6] who found no significant difference whatsoever between their two groups of students. The same baffling reversal of conclusions has been obtained with many other early investigations which appeared to demonstrate the greater effectiveness of discovery learning. Recently G. Herman[7] has analysed over 30 such studies and, in the majority of cases, has found only insignificant differences in success ratings between these two opposing teaching methods. The only conclusions to be drawn from these statistics seem to be the difficult nature of this sort of research and that the successes once widely predicted for the discovery method are still not proven.

However, it may be that the speed and accuracy of learning a specific topic and its use in problem-solving, commodities which are relatively easy to measure, are not relevant criteria for the success or failure of the method as education. It has always been obvious that discovery is not the most economical approach to learning: it must take more time to absorb information in this way than by traditional methods where careful, precise instruction is the prelude to all practical work. It may well not even be the most efficient method. The advantages most often boasted for the discovery method are less testable but not, perhaps, less real – increased enjoyment and an enhanced interest in science as a whole. These more elusive, affective factors have also been the subject of research. Whole classes of pupils nurtured on discovery methods have been extensively quizzed on their liking and general attitude towards science. Once again, very little positive evidence has emerged. Indeed reports[8] on the attitudes towards physics among some students following the Australian version of the PSSC course indicate an actual decline of interest.

## The Case Against It

The other recent line of attack on the use of discovery methods, which is led by David Ausubel from the Education Department of the University of Illinois, is theoretical rather than analytical. As early as 1961, when the full tide of fashion in education was running for discovery and against verbal instruction, he published an article 'In Defense of Verbal Learning'. The point that Ausubel made so forcibly is very simple – it is the *meaningfulness* of the learning structure, whether verbal or practical, which brings about cognitive growth in children. From this, he argued, and continues to do so, that where practical work is discovery-orientated to the point of encouraging an unstructured encounter between pupil and apparatus before instruction and the introduction of meaningful scientific ideas, it does not easily bring about learning. Carefully structured verbal 'reception' instruction, on the other hand, can be readily understood and fitted into the individual's existing body of knowledge where initial inquiry or simple trial and error are both unnecessarily difficult and time-consuming.

> The learner must rearrange a given array of information, integrate it with existing cognitive structure, and reorganise or transform the integrated combination in such a way as to create a desired end product or discover a missing means-end relationship. After this phase is completed, the discovered content is internalised just as in reception learning.[9]

## On Calling a Spade a Spade

Although the verdict of this chapter on the validity of class 'discovery' is emphatic and negative, this in no way detracts from the value of experiment. It would be sad indeed if this argument were wrongly construed as a call to return to 'chalk and talk'. The teaching experiment is a tool of the greatest importance and even delicacy; it does nothing but harm to the integrity of our lessons, and to science itself, to dress it in false colours. Pure, open-ended discovery, as we have seen, must, by its very nature, be an individual and spontaneous activity. By contrast, when twenty to thirty children are being taught in the school laboratory, their experiments can serve a wide variety of useful purposes, some of which we will attempt to analyse in later sections.

Unfortunately there are still many teachers, influenced perhaps by a naïve belief in the nature of scientific research, who feel bound to pretend that every experiment is in the nature of a startling revelation. From the children's point of view many experiments are so obviously designed to show a particular effect that the expected result is, at least in outline, transparently obvious before the experiment has begun. Suppose two containers are set up, one with ingredient X and one without it, then a title 'To *discover if* X affects. . .' will deceive no one. As the children say, 'You wouldn't have done it like that, Miss, if there wasn't a difference!' Such an experiment is more honestly entitled 'To *show* the effect that X has upon. . .' Yet still the experiment can be important, fresh and interesting. There are practical details to be discussed, the magnitude of the effect to be observed and illustrative power to be given to the preceding theory.

Biological experiments fall most often into this category because the presence of the control is so important,[10] thereby giving away the whole show beforehand. It is also in biological experiments that the most careful and detailed observation needs to be made, often over a considerable period of time; it is comparatively rare for there to be an instant and unequivocal result. The effect of falsely pretending that such an activity is a voyage of discovery into totally unchartered waters is not only counter-productive, it can also be acutely embarrassing to the struggling teacher, as is shown by the following true story.

Teacher: 'Today we are going to do an experiment to find out if peas need to respire during germination. Here is the apparatus.'

Pupil: 'Please Miss, we don't need to do the experiment because you have already told us that they do need oxygen.'

Teacher: 'But how do you know I'm right? You have got to prove it, haven't you?'

After a week they examine the apparatus again. Germination has

Figure 3.4

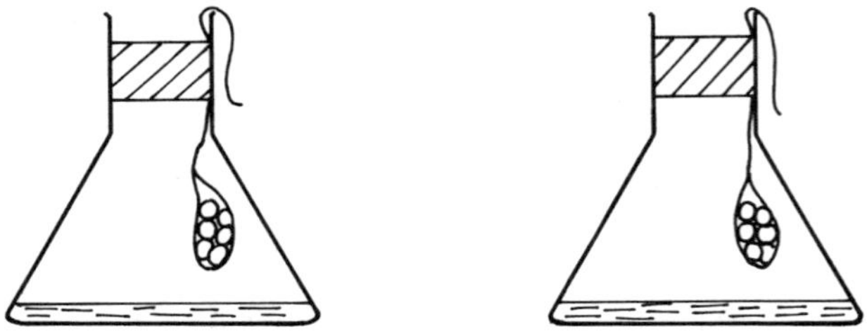


occurred in both flasks and the anaerobic nature of the growth in flask B has not yet become apparent. (If it has, the growth will appear *more* prolific.) This would require at least another week of continuous observation. Meanwhile the unimpressed pupil makes the predictable response:

> Pupil: 'Oh I see, we did have to do the experiment to prove that you were *wrong*! Peas don't need oxygen during germination.'
> Teacher: 'No, no, they do. It's just the experiment that hasn't shown it.'
> Pupil: 'Shall we write down that the experiment went wrong?'

Such smart and knowing pupils lurk behind almost every bench. Occasional contaminated reagents and broken electrical leads will give them opportunity enough to score without handing it to them on a plate!

At an earlier stage it is not at all obvious to children that experimental results can be predicted from theory and that science is expected to 'make sense'. The predictive experiment is therefore very important and requires a chapter of its own.[11] On the other hand, there are a few experiments which are so simple and reliable that we can depend upon them to generate surprise by running counter to the effect that the children themselves believe will occur. These expectations are so hopelessly wrong that the shock effect of the experiment is a welcome ally in a frontal assault. This much at least we can borrow from our analysis of discovery.

Once again a mis-worded title to the experiment can rob the result of its valuable impact. Most young children (and not a few adults) believe

(1) that light bulbs use up electricity; and
(2) that sourness or acidity can be cured by sugar.

The relevant experiments are most often called 'To find out about the current flowing in a circuit' or 'To discover the effect of mixing an alkali with an acid'. Since these are directed experiments, we need not retain the opening verb. Much more mileage can be coaxed from these activities if they are presented as

(1) measuring the amount of current that goes into and out of a light bulb; and
(2) neutralising lemon juice with sugar and with indigestion tablets.[12]

Sometimes pupils can hardly believe their own results, but these are almost foolproof experiments and massive corroboration from the rest of the class, as well as some cross-examination of the teacher, compels belief. These experiments are by no means pure discovery, but the element of surprise, which children relish so much in game and story, goes far towards making the results memorable for them.

**Notes**

1. Carl Bereiter, in *Farewell to Schools???*
2. Physical Science Study Curriculum.
3. Chapter 4.
4. Chapter 5.
5. L. Szekely, 'Productive Processes in Learning and Thinking', *Acta Psychologica*, vii (1950).
6. I. Maltzman, E. Eisman and L. Brooks, 'Some Relationships between Methods of Instruction, Personal Variables and Problem Solving Behaviour', *Journal of Educational Psychology* (1956).
7. G. Herman, 'Learning by Discovery: A Critical Review of Studies', *Journal of Experimental Education* (1969).
8. Gardner, 'Deterioration of High School Students' Attitudes to Physics', *Nature* (1974).
9. D.P. Ausubel, 'In Defense of Verbal Learning', *Education Theory*, 11, (1961).
10. Chapter 8.
11. Chapter 7.
12. Nuffield Combined Science, Section I.8b.

# 4 INDIVIDUALS AND UNTAUGHT LEARNING

## Difficulties of Class Teaching

A school may be a collection of individuals, even though headmasters monotonously and mystically insist during morning assemblies that it is much more than this, but teaching a class is certainly vastly different from watching over a series of voluntary hobby activities as described in the second chapter. From the shy five-year-old entering a primary school to the committed adolescent of the sixth form, every pupil is slotted amiably or protesting into a class. For the teacher there are now the many-headed mob reactions as well as the variously coloured characteristics of the individuals composing it. Whether the subject to be taught is English literature or integrated science, the perennial problem of how to direct the teaching, towards the social entity as a whole or towards the children as individuals, can never be definitively resolved. The topic is polemic and explosive, with fireworks contributed by psychologists, parents and politicians as well as by the more pressing demands of discipline and the syllabus – however vague. It would not be useful to tread a careful path that skirts all these issues, but it is essential to keep in view the objective, which is science teaching, and the venue, which is the school laboratory.

Whether the class being taught is labelled 'mixed ability' in an ideologically pure comprehensive, or is a sifted, selected group within an ability-streamed school, real uniformity is always out of the question. It must be superfluous to labour the point that preferences, behaviour, interests and speed of learning will always range over a wide spectrum. This inescapable diversity raises problems which loom very large and menacing in the actual environment of the teaching laboratory. The first of these is obvious from the outset. How can each child be taught by the intimate step-by-step method which mirrors his own characteristic learning approach? Just outlining the problem which has been selected for the day might well require half a dozen different phrasings and analogies, jokes, hints and incentives in order to appeal to the variety of tastes present and galvanise the rows of moody upturned faces into the lineaments of eager and involved participation. That is the ideal of teaching, hard for any, and perhaps for the beginner almost prohibitively difficult. The other problem is the reverse of the coin. Because what is interesting to one pupil may be pointless and boring to another, especially when there is a wide spread of abilities, there is a serious risk of losing the attention of a few disruptive pupils at the start of a lesson which necessitates some stern measures to re-establish good order and so loses the easy flow of explanation. These are very real dilemmas in the daily craft of teaching where it is inexorably demanded that one adult should somehow contrive to hold the reins of attention of thirty lively youngsters, charming each and every one beyond the simple goal of good behaviour into the busy realm of happy learning. Pursuing such a hard ideal can be sadly expensive of the spirit and so, inevitably, an alternative way is sought.

## The Arrival of the Worksheet

The method came to hand almost by accident. Once the science textbook had been cast out as a vehicle of middle-school teaching by the discovery-orientated courses of the 1960s, the pupils were left with nothing but oral instructions for experimental work. The early Nuffield courses were satisfied with this, and were, indeed, very teacher-based. All the children possessed were question books from which to do their homework. These courses were explicitly intended for the upper-ability grammar schools, but as soon as courses for mixed-ability classes were envisaged, precise 'activity' sheets became a necessity so that those who were slow to comprehend instructions could pore over them in written or pictorial form instead of requiring endless verbal repetition from the teacher.

Not only did the Nuffield Combined and the Scottish Integrated Science courses publish their own instruction sheets, schoolteachers were themselves quick to pick up the cue. It is now common practice for schools to duplicate their own work cards, and they take great pride in this activity, exhibiting them to neighbouring schools and even publishing them for a wider audience. This is an important new departure in school teaching which cannot be over-emphasised.[1] By its operation the impetus in education can be seen to be moving away from the university departments of science and education which launched the original innovations, and into the schools themselves, where these basic teaching and curriculum tools are fashioned to suit the local needs. These sheets not only replace the more expensive junior textbook, but also ensure uniformity and continuity of teaching within a large school department with changing or absent staff.

Such worksheets profoundly influence the sequence of learning. The received method of teaching began with a discussion of the topic which forced the format of the lesson into a frontal teacher-led exposition and, as we shall see, there are strong pressures, both ideological and practical, against such an approach. On the other hand, if the nature of the problem is reduced in importance so that the pupils begin by being launched straight into a worksheet full of precise instructions for an experiment, the lesson will wear a very different aspect. This is the common method, action-loaded 'discovery', highly regimented and, no doubt, a travesty of what Schwab and Bruner intended. Nevertheless it has its conveniences.

## The Format

It must not be supposed that the conventional worksheet is the pure-bred descendant of the time-honoured laboratory manual. Interspersed among its purely practical instructions are telling questions such as, 'What substance are we testing for?' or 'When before have you met such an action-at-a-distance?' with a space for the mandatory answer. Every experimental result is capped by the inevitable question 'What does this prove?' couched in some helpful and suggestive phraseology. For example, a worksheet designed to teach photosynthesis at second- or third-year level is apt to begin abruptly with the bare title 'Examining Photosynthesis part 1' and no more statement of a problem than 'You are provided with two plants one of which has been grown in the light and the other which has been kept in the dark for the last few days.' Only after copious instructions for the step-by-step procedure involved, cutting out pieces of leaf, breaking open the cells in boiling water, seething in alcohol (complete with safety precautions about the proximity of flames), placing on a white tile and covering with drops of iodine solution, only then is the nature of the 'problem' revealed. It seems a little late! Hitherto the author's anxiety to preserve some vestigial element of discovery in the experiment had transformed it into little more than a practical obstacle race, so now the importance of the exercise must be hammered home in retrospect. Not only 'What colour did the leaf turn?' and 'What substance must be present in the leaf?' but also 'In which plant, that grown in the light or that grown in the dark was starch formed?' and 'What does this tell us about the connection between light and the production of food in a plant?' This is frog-marched discovery at a very far remove from the 'research' type of project envisaged by Armstrong or Schwab. The worksheet too has been transformed. It is now not so much a guide to experiment as a completely packaged lesson, even a programmed teaching aid in keeping with modern psychological precept.

**The Range of Ability**

The first hurdle that the worksheet meets is the need to be easily read and understood. It is quite straightforward to ensure that the 'reading age' of the material is some two to three years lower than the age of the class, and where the writing is exclusively concerned with commands and instructions, this simplicity of vocabulary is no great drawback. It does mean that the worksheet will be a poor vehicle for the introduction of theory of any complexity, but we shall see that there are other profound reasons why the worksheet is unsuitable for this. Some skill in the wording is required, since even the simplest instructions can be misunderstood and where potentially dangerous apparatus is to be used most teachers will always play safe and themselves read out, explain and demonstrate the equipment in operation before the children begin. Finally there will, sadly, be those in every class who cannot read effectively enough for even the most basic activity sheet. Pictures help here, but a sympathetic adult helps even more, for such immature children often need extra emotional support every bit as much as they need guidance in reading. The printed word is not their natural medium and a spoilt experiment teaches nothing whatsoever.

Inevitably the rate of working raises another problem. What should be done for those pupils who finish long before the others? This deceptively practical question turns out to be an ideological point of departure. To those who are wedded to the anti-elitist view there is no work intrinsically more suited to those of 'higher intelligence' and they see no more difference here than in *speed* of working and understanding. According to the Carroll[2] theory of classroom learning there are no variations in specific aptitudes and hence there is no need to provide the kind of extra challenge that those involved in the National Association for Gifted Children, for example, might advocate. No one, of course, wants such children to become bored or even inactive, so the choice becomes one of either letting them race far ahead of the field along the same track which is organisationally awkward, or of providing extra sheets to keep them busy on similar experiments. This way the whole class can be kept roughly in step and examined on the same material. The radically different approach would be to provide work of a more abstract and formal nature, supported perhaps by a belief in the cognitive development theory of Jean Piaget.

Can the latter view be accommodated within an 'extension' series of worksheets like those provided by 'Insight', the new science course for 11- to 13-year-olds which has been produced by London's ILEA? The majority of the extra work cards do not introduce any further subtlety of thought and those teachers who struggled on their own with this problem have also found that the format of the worksheet, although splendid for practical instructions and 'question and answer' sessions, is not satisfactory for the teaching of theory, a point to which we will return in the next section.

Nevertheless the worksheet remains probably the only viable method of mixed-ability teaching, and those who are confronted by the problems raised above can sometimes be heard publicly consoling themselves with the comforting statistic that hardly two pupils in an average class of thirty 13-year-olds will have advanced to even the lowest rung of the formal reasoning ladder. Piaget himself has written that it is only by the 'use of active methods which give broad scope to the *spontaneous* research of the adolescent'[3] that pupils can develop to this searching, independent phase of thinking. Whether mandatory worksheet experiment or teacher instruction is more contrary to such a dictum is hard to judge!

The process that began so innocently as a list of instructions for experiment and a table for the results to be entered has been transmuted by the philosophy of de-streaming. The worksheet now wears the mortar-board! Almost every function of the lesson-giving teacher can be translated into a written analogue: testing previous knowledge, giving information, setting problems and even asking for the invention of an experimental test are all to be found on one worksheet or another. The response may vary from one extreme, where no more than half a dozen words are written into the appropriate gaps in the printed sentences, to instructions to 'answer in complete sentences' or 'Describe the whole experiment in your books.' Under the influence of the Bullock Report on language across the curriculum within the school the latter method is becoming the more common.

To complete the transformation even the teacher's hoary jokes have reappeared as trendy cartoons and the authority of his personality as a visual stimulus-response code on the worksheet:

When you see this sign →  – WRITE.

When you see this sign → 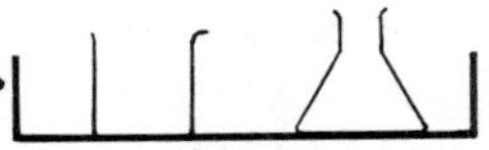 – COLLECT YOUR APPARATUS.

A coloured background means COPY OUT THIS PASSAGE.

Words printed in blue like this 'DISTILLATION' must be copied into your SCIENTIFIC DICTIONARIES.

Evidently programmed learning has arrived.

**The History of Programmed Learning**

The methods of B.F. Skinner, the 'Father' of programmed learning, can be variously regarded as inhumanly behaviouristic or as an individualistic aid to good teaching along the lines which have always been used by patient human tutors. Skinner's original research involved training pigeons to perform complicated tasks, like treading out a complete figure of eight, by reinforcing with a gift of food even the smallest step or movement in the right direction. In this way he built up behaviour far more intricate than that which mere chance, like pecking at a particular disc, could ever have achieved by rewarding the end product. He noticed in this early work that having to extinguish a wrong response produced a disturbing emotional reaction and this, together with the technique of advancing in very small steps, became the two basic ideas that he subsequently brought into the field of human learning. He pointed out that 'aversive reinforcement' – punishment for the wrong answer – which had been the major and often painful inducement to learn for so long, was now no longer so easily available, as well as being emotionally detrimental to the process. Instead of bemoaning its disappearance and inventing newer methods of discipline, we should use a far more effective weapon, the *positive reinforcement* of each correct response, however small. Curiously enough, it was these 'soft' approaches to teaching, avoiding failure so that the student should enjoy only 'a fortunate history of successes' and continually rewarding each right answer, that earned Skinner such opposition from the so-called 'humanistic' school.

In the light of our present knowledge a school system must be called a failure if it cannot induce students to learn except by threatening them for not learning. That this has always been standard pattern

> simply emphasises the importance of modern techniques.[4]

Nevertheless, Skinner was far from insisting that his method, as opposed to his technical hardware, teaching machines and programmed texts, was in any sense sensational or new. Every parent who has ever sat with a child in his lap reinforcing with kisses and delighted cries of 'Good girl!' or 'Who's a clever boy?' the infant's first attempts at speech or drawing should recognise in the writings of this arch-protagonist of the cold teaching-machine some glimpses of that first Eden of all our learning. Discussion of techniques will be more controversial, but first let us salute an analysis of teaching which is simple, affirmative and familiar.

> Men learn from each other without being taught. A man may once have learned to use a digging stick by watching someone else use one, but the digger was not therefore a teacher. It was only when the increased effectiveness of the learner became important to the digger that he became a teacher and changed his behaviour in order to facilitate learning – moving more slowly or exaggerating his movements so that they could be more easily imitated, repeating some part of an action until it could be successfully copied, reinforcing good digging with signs of approval, arranging roots so that they could be more easily dug.[5]

A brief survey of the history of the gadgetry of programmed learning or 'operant conditioning' is worthwhile, not because it has found a firm place in the classroom – teaching machines have been a rare[6] commodity in the English school – but because its indirect influence has been so pervasive. The earliest machines, like all their kind, arose from the effort to remove some of the drudgery from daily work. In the 1920s S. Pressey of Ohio University devised a test-scoring machine in which a pupil could pull down an appropriate lever to indicate the answer to a simple question. If correct, the machine would record his score and present the next question. Obviously several answers had to be available in the now familiar layout of the multiple choice system. At the time there was little response to this innovation and Pressey himself was despondent at the apparent failure of his work. In retrospect we can now see that his idea needed only the pressure of an increased school population and the advent of the microcomputer to force some kind of mechanical test-scoring system on all schools, at least during examination periods.

Subsequently Pressey made another claim for his machine. Since the next question could not be posed until the first had been correctly answered, the pupils were forced to find the right answer and 'rewarded' when they did so by progress to the next frame. In this way the machine indicated whether an answer was right or wrong and Pressey claimed that it not only marked tests, it also 'taught' the children in much the same way that a teacher would. His claims went further. Quoting the 'Law of Effect' and the 'Law of Recency' formulated by Thorndyke and others at the turn of the century, he insisted that this immediate mechanical reaction must be superior to many forms of traditional teaching where correcting and reinforcing the answer were delayed, sometimes for several days, while the teacher marked the pupils' books. One of his associates tested out Pressey's method on a group of first-year university students who were allowed to test themselves on the machine twice a week over a period of several months. The examination results at the end of the semester showed that these students did far better than those who had not had the advantage of such tests. Pressey's followers were jubilant; this research seemed to prove that a machine which could correct mistakes immediately was an effective and successful teacher. The educational world remained unimpressed, whether due to lethargy, lack of finance, revulsion from mechanical teaching or scepticism in the face of this specious argument, one cannot tell.

Skinner's machines, which arrived some thirty years later, were overtly designed to teach and not to test. They took on another aspect of the teacher's function by presenting new information, albeit in very small doses, and there was no attempt to probe the learning achieved by offering wrong answers in a multiple choice layout. In teaching, as opposed to testing, it would be obvious folly to suggest an incorrect solution. Instead the pupil was usually required to fill in a space in a sentence with the appropriate word which he had just been given in the previous snippet of teaching. In this way it was hoped that the pupil would be perpetually rewarded and reinforced by his easy success – like the waltzing pigeons! Indeed Skinner insisted that any question which proved to have an error rating of 5 per cent or more should be rejected in favour of an easier, more suggestive one.

The early 1960s were a very good time for educational experiment in America, as has already been indicated, and Skinner's machines were adopted quite widely. In this country they have been less successful, but the Skinner-type question is undoubtedly here to stay. It is to be found in almost every worksheet and in every kind of lesson. There was a time when all questions addressed to pupils were designed to test their knowledge and problem-solving ability. This is no longer the case. The object of questions asked on the modern worksheet is to teach by eliciting only the right answer.

> We have found that only the plant grown in light formed starch in its leaves. Fill in the right word in the space below:
> Green plants need. . . . . . . . . . . . . . for the production of starch.

Obviously it is quite possible to produce a whole programmed worksheet of this kind with snippets of knowledge followed by sentences with missing words which nudge the pupil into filling in a space before accepting the next information-pill. This method owes a lot to Skinner but, in the absence of an automatic and immediate scoring device, it lacks a vital ingredient. If there is no signal of success, there is neither reinforcement nor reward for the student's effort. There are a few books[7] available in which the action of turning the page reveals the correct answer but, in my experience, few children or adults can resist such an inducement to cheat! Teaching machines, on the other hand, are not only cheat-proof, they are fun to operate and introduce an element of competition – 'beating the robot'. However, the novelty soon wears off and in tests which compared the learning rates of children using the same programme presented by machine and by text, very little difference could be discerned. Good school programmes are themselves in very short supply, but before we bemoan their lack with too much desperation it has to be shown that they really are more effective than the human teacher with human reinforcement.

## Research Studies on Programmed Learning

An extensive experiment was carried out in eight Glasgow secondary schools during 1963 involving more than 600 first-year pupils. The subject chosen was an introduction to algebra lasting 15 weeks, and in each school there were two matched groups using programmed texts and 'a real live teacher' respectively. At the end of the period both groups were given the same test and in seven out of the eight schools the programmed group scored significantly *worse* than the traditionally taught control group. Another test was given some three months later to gauge the retention of the material and, once again, the programme-taught children achieved a lower score than those in the control group.

Mathematics would seem to have been a very appropriate subject to be taught by this method and even so the results, as we have seen, were disappointing. No such large-scale tests have been conducted with scientific subjects. The nearest approach has been a similar piece of research on the operation of a geography programme concerning the earth's rotation, day and night, latitude and longitude – all topics very close to science. Here, as would be the case in a science programme, the printed page and flat diagrams had to compete with three-dimensional working demonstrations available to the traditionally taught groups. Presumably a visual demonstration is considered too large an item of information and too non-verbal to be included in the programme technique. It is not surprising to find that in this experiment the programmed group did very much worse than the control group.[8, 9] Admittedly these results are unlikely to have much effect on the threatened microcomputer revolution in schools. The programmers, now computer technicians without a learning theory, have advanced from the design of TV games to educational 'software'. It remains to be seen if the children of the 1980's respond any differently to those of the 1960's.

## Social Influences at Work

Yet, despite this discouraging research, programmed worksheets and textbooks are still advancing apace, which is a sure sign that the forces at work in educational circles are other than those of simple efficiency. Some of these are related to an ideological stand on mixed ability teaching, some to a tighter control on the method and substance of the teaching. Another influence stems from out-of-school forces which are frankly antagonistic to the whole concept of the teaching teacher. Ever since the days of Dewey and the advent of the pupils' own project system, there has been a swing away from person-to-person teaching. The reasons for this certainly included a strong urge to encourage independent learning, which can hardly be considered relevant to the detailed step-by-step instructions of the modern worksheet. There was also a deep suspicion of heavy teacher domination and indoctrination from which the pupils needed to be 'liberated'.

To see the programmed method of education raised to its final peak of importance we have to look to America or Sweden. Here, it is claimed, is the goal of individual prescribed instruction, every child having his own programme to which he has been machine-matched, like a computer dating service. At the expense of any element of individual choice the education has finally been 'individualised' – a telling paradox! The teacher is still there, but heavily disguised as a cross between a filing clerk and an overworked doctor. His job is to analyse and correlate the results of tests, to diagnose the next learning objective and to prescribe the appropriate treatment. To those who design such schemes (SMILE, PLAN, IPI or IMU) they seem wonderfully child-centred and able to be tailored to fit any figure.

> The educational researcher. . .envisages the teacher as a physician who, after having made his individual diagnosis, writes out the best prescription for a pupil and tells him to have it filled out by a pharmacy of tried and tested didactic alternatives.[10]

There is another pressure from outside the school that wishes to replace the teacher by 'instruction learning'. The changing technology and employment has made our society 'information rich and action poor', a phrase of J.S. Coleman which he uses to advocate training the pupil's own action and initiative to cope with the flow of information. The Swedish educationalist T. Husen predicts a cycle of recurrent education throughout life for which school is merely the initial preparation: 'The central mission of the school will be to teach the pupils how to learn, to train them to assimilate new knowledge on their own.'[11] Whether the objective is further education or retraining for new 'information' work, this attitude implies direct encouragement for the worksheet method so that training pupils to follow instructions becomes a goal in its own right.

Substituting the word *training* for *teaching* implies more than mere distrust for the power and prestige of a teacher. It is the call of society's real concern for its future citizens. Are the schools preparing them properly for the circumstances of their adult lives? Can they follow written directions, work out accounts and accept retraining when the tides of unemployment submerge the older types of jobs? No one would quarrel with such concern nor wish to resurrect, in some new guise, the sterile academicism of the old public school with its exclusive, remote study of the classics. However, there is another face to the coin. The motives of the Victorian foundry master who called for less education and more training for the children's 'proper station' were only too gruesomely obvious. Can we strike a better balance between *training* our pupils with socially orientated courses for this post-industrial society whose store of available employment becomes steadily more information-rich, and *teaching* them as the humanists would have us do it, so that the right windows are opened for each child to enrich his life with colours more vivid than the sombre hues of the 'job-skills'?

Experience gained from the use of programmed material in industry and the armed forces shows that it can indeed be excellent for short training courses. Educational research statistics from the classroom and the laboratory, as we have seen, are less unequivocal both because the objectives of science teaching are wider and because the reactions of our children are more varied. These aspects need more consideration.

**Is Your Teacher Really Necessary?**

One obvious problem with programmed activities might be the absence of the Skinnerian signal of reinforcement. Children cannot be starved like animals in training sessions so that they can be rewarded by food. Continual gifts of sweets would produce an outcry from nutritionists and dentists alike! However, they can be rewarded by fun, by praise, and by approving smiles from a teacher. In adulthood it is easy to forget the old respect we used to feel and dwell in memory only on the smarts and humiliations received in failure and childish impotence. Within the school situation praise, even from a feared or disliked figure, is almost always strengthening and delightful. It contains an element of the essential parental love and approval syndrome on which all happy children are nourished. 'Liberation' is a cry that originates outside the school. In the pioneering attempts to produce completely individualised learning within schools the teacher still needs to be very much in evidence, not only to ensure safety in the laboratory but also to provide the essential ingredients of personal contact and encouragement.

Programmed mathematics courses have a longer history than do the new 'individualised' science worksheets which are just coming hot off the school duplicators, and their use has not been entirely happy. They differ from science courses in two respects. In the first place, each child is expected to work his mathematics course completely on his own, whereas science is more usually aimed at an experimental group of two or three children.[12] Secondly, a stern withdrawal of help in the interests of encouraging self-reliance is more readily practised outside the laboratory where there is less chance of experimental mishap! This may even dilute the rigid purity of the programmed approach to some advantage by avoiding the sullen boredom engendered by the isolation of the blinkered workcard path. Children are gregarious creatures, their happiest lessons are full of interaction, co-operation and competition.

Science too is a social activity and where it can be matched by sociable learning both the children and the lesson stand to be enriched by the mixture.

The children will need help. Again and again in a programme or a worksheet that is couched in language as simple as we can make it, some children meet an impasse. It may be a word or a phrase which is not understood or a new idea that makes no contact with that child's way of thinking. Now the worksheet's promised liberation could become an open prison to the isolated child. Branching programmes, still a rarity in the laboratory, might re-teach the idea in another way, but who is to hear the child's own faltering explanation of what seems so difficult or watch his face as some other lines of explanation are attempted? Even the most modern teaching machines admit this possibility by the inclusion of a 'HELP' button designed to attract the teacher's attention by a trendy flashing light in place of the mute puzzled stare or appealing cry of 'Miss!'

With every child at a different point in the course even the giving of help assumes a new complexity. No general instructions can be issued to the whole class and each pair of pupils may need an individual guiding chat from the teacher. At their best these classes can seem like the golden mixture of personal tutoring and independent learning that every teacher dreams of. This requires a great measure of skill and, unfortunately, it is not the only scenario. With doses of help given in busy rotation there can also be idle waiting, irritation and misbehaviour, frustrated calling out or sturdy continuation along totally wrong lines. The originators of these new mixed-ability methods of teaching are almost universally enthusiastic about their results[13] and it is far too early to make any useful overall judgements. The well-known 'Hawthorne Effect' in trial schools always stimulates an atmosphere of euphoria which limits vision, but it does seem that the greater the range of ability within a class, the greater also are the calls upon the administrating teacher.

The correct Skinnerian response to this difficulty is to make the learning steps smaller and easier so that less help is required by the pupils. This is being done with some success, but also with an inevitable slowing down in the rate of covering a syllabus. One such individualised school course now includes no more than the measuring of length and volume, the main groups for classification of animals, and separation by filtration and evaporation in *two whole years* of integrated science with four teaching periods per week. Clearly there is a balance-sheet of loss as well as gain in this method.

**Better on their Own**

We seem to have cast the teacher in the role of administrator, tutor and encourager. It all sounds a little too like the sentiment of 'guide, philosopher and friend' and it is as well to remember that although such beautiful gifts may be offered, they may also be rejected. Among the children in our schools are the angry and quarrelsome, the rebellious and sullen, the aggressive and destructive. Some of the remedial methods of using science teaching in these real and depressing circumstances are discussed in Chapter 6; how then does the individualised approach match these problems? The two most awkward characteristics of these troublesome children are their difficulty in accepting an 'authority' figure and, equally, their quarrelsomeness with each other. They react angrily to any teacher confrontation, hate all discipline imposed by adults and are touchy to the point of fighting with each other over an argument. To such pupils the low profile of the adult teacher and the chance to be doing, or not doing, in their own unpressured time can be positively soothing. These children are often better working alone and the individualised worksheet which prevents sociable co-operation to the detriment of some class work can also, by the same token, limit inter-pupil argument and friction. Some teachers are convinced that its use has improved the worst of the behaviour nightmares in science lessons.

**Person or Paper?**

The worksheets are the products of teachers but, as we have seen, they cannot supplant their creators like some computerised robot from science fiction! They have certain advantages over the unaided human teacher, but without his organisation and encouragement they lie like so much litter on the bench. Consider the completely individualised course, where the lesson notes of the most gifted teachers have been translated into clear comprehensible instructions and information for the children, when they in turn enjoy the system and are familiar with its expectations, when the teacher has adapted to his new back-up tutorial role – are there any valuable teaching components which have been lost in the transition? Emotive cries of 'inhuman', 'behaviouristic' will not do. The human teacher has not been banished from the laboratory, nor from contact with the children: only the method of learning has changed.

In the traditional system the teacher had the possibility of adapting his methods as occasion demanded. Many had their own favourite teaching dodges and the best had a whole armoury of alternatives. The interaction between teacher and class can be a complex prompting situation with a host of ideas and suggestions being offered from both sides until enough are grabbed out of the air for an experiment to be arranged. That is the ideal. Take, for example, a first-year class which is beginning to grapple with the concept of volume. One worksheet may begin with fluid in a measuring cylinder introducing the unit '$cm^3$' as a relation of the more familiar pint or gallon. Another will start with the measuring of large blocks of wood or metal. Eventually the two approaches must overlap; the cubes must 'flow' into measuring cylinders or the liquids must become comparable to solid shapes. Neither method is intrinsically superior to the other, but circumstances can make one of the approaches obviously wrong for the occasion. If the pupils themselves suggest one method it would be absurd to thrust the other at them. It can happen that a whole class jibs at the very idea of three-dimensional measurement. This may be a restless inattentive group or just the end of Friday afternoon, and as the teacher struggles in the fog of incomprehension some new approach may suggest itself – it might involve cutting up lumps of plasticine into cubes with a blunt knife or building different shapes out of wooden blocks. And so he innovates. The programmed worksheet can be tested and remodelled over a trial period but, in the end, it is fixed into the scheme to be encountered in identical form by each pupil in every class.

The diagnostic teacher uses his empathy to sense the understanding of the whole class through the medium of discussion. Where simple practical skills are being taught, as is so often the case in the early years of secondary school, the lack of this discussion and variety may be little felt. However, when concept-based experiment is the object, prior discussion becomes quite essential (Chapter 7). This may not preclude the use of worksheets but it does mean that the whole of the class must be at the same point in the work so that they may all contribute to the discussion and profit from it.

There are many occasions when class discussion can arise quite naturally out of a pupil's public question. Not all questions are equally useful, some like the advanced and idiosyncratic 'Why should hot atoms vibrate?' or the humiliatingly stupid 'How do you spell "wet"?' are much better answered in private. But there are fine lateral questions that everyone can enjoy together, like 'Are we descended from dinosaurs?', which enliven dull sessions on the classification of animals quite superbly.

There is also another, more elusive quality of frontal human teaching that the flat worksheet cannot hope to emulate, which may appear trivial, but has very basic learning value. It is a matter of body movement and explanation. Once again we are concerned with the more conceptual character of science teaching. It is shown in Chapter 5 that a surprising number of these concepts are closely related to everyday sensorimotor activities and before they can be usefully internalised the latent experience must be evoked and remembered. Often this necessitates gestures, chat and even simple acting. Many examples could be given, from the energy needed for a molecule to evaporate out of a liquid to the mechanism of breathing.

The writer of one apparently successful individualised course has admitted that even the elementary concept of volume has had to be dropped from her two-year science course because of a failure of understanding. Now this is a topic that has fascinated educational psychologists since the earliest work of Jean Piaget and it is profusely documented. It is well known that he found many eight- and nine-year-old children who seemed unable to recognise the conservation of volume even when the liquid was poured from one container to another in front of their eyes.[14] More recent work has indicated that it may have been only the form of the question and the limits of the child's vocabulary giving the impression that these young children did not recognise this quantity. Certainly no one has ever suggested that 11- or 12-year-olds would be incapable of understanding so simple a concept. This curious failure of individualised learning demands another explanation. Could it be that this new word 'volume' required a few encompassing human gestures to link it with the simple 'all-round bigness' that every child has experienced? Whenever science learning requires imaginative prompting rather than step-by-step instructions the worksheet with its careful sentences must be at a severe disadvantage. It needs a *human* teacher to act out *human* reactions of feeling or movement, which will be among the main topics of the next section.

## Notes

1. *Making Inset Work,* Department of Education and Science Discussion Paper, (HMSO, London, 1978).
2. John B. Carroll, 'A Model of School Learning', *Teachers' College Record,* 64 (1863).
3. J. Piaget, 'To Understand is to Invent', *The Future of Education* (Penguin,

Harmondsworth, 1979).

4. B. Skinner, Review Lecture at the Royal Society, 1964.

5. B. Skinner, *The Technology of Teaching* (Appleton, New York, 1968).

6. There are already signs that the cheap microcomputer may make a new bid to fill this vacancy.

7. For example, D. Reid and P. Booth, *Biology for the Individual* (Heinemann Educational, London, 1971).

8. Swansea, 1962. Spolton report in *Univ. Coll. Fac. of Education Journal.*

9. Larkind and Leith, *Programming 1963* (Sheffield University), or Herringshaw and Hunter, *Educational Review* (Birmingham University) (1964).

10. T. Husen, *The Learning Society* (Methuen, London, 1974).

11. Ibid.

12. The new Advanced level independent learning schemes, such as APIL, are often worked individually. Although it is too soon to judge these, there are already indications of pupil dissatisfaction with the isolation produced.

13. L. Sturges, *Non-Streamed Science*, ASE Study Series No. 8 (Hatfield ASE, 1975).

14. Jean Piaget, *The Child's Conception of the World* (Routledge, London, 1929).

# 5 IMAGES AND CONCEPTS

## Seeing Past the Apparatus

The basic philosophy of the discovery method for teaching science seems to be that the children should learn directly from what they observe in experiments. One might usurp the Churchillian thunder, 'Give them the tools and they will get on with the job!' But then children have neither the dedication nor the expertise of wartime soldiers; before battle commences some of the most delicate and fascinating tasks in the whole art of teaching will have to be attempted.

Whatever the tools that are put into the eager hands of our children, bunsen burners, ammeters, hamsters or microscopes, it will not be about them that the experiment will be concerned. Its objective could be heat, current, behaviour or plant cell structure – intangible mysteries all, whose only lines of communication with our understanding must be the solid apparatus of the laboratory. Since these abstract concepts are to be the main actors in the forthcoming experiment, then they and not the items of equipment must be followed in the mind's eye of the child throughout all the hurly-burly of activity. Not only what makes things happen but also what is actually perceived by the pupils, and even the order in which the experiment is performed and carried out is heavily dependent upon the prior recognition of this abstract concept.

## What Makes it Happen?

The effect of misidentifying the agent causing an observed experimental happening has often been all-important, even for mature research scientists. The chequered history of Brownian motion provides an excellent example. The story of how the botanist Robert Brown first observed the curious jogging motion of pollen grains in water is well known. He felt quite confident of his ability to interpret this movement from the start of his research. Pollen held for him the vital principle of vegetable life and where there was life he could readily accept that there would also be movement. He not only saw the 'active molecules' oscillating under their own power, but even changes of shape, swellings and contractions, produced, one must assume, by the rotation of irregularly shaped particles. All this he took as irrefutable evidence of the basic life-force, rather in the same way that we see the formidable pulsations of the living heart when laid bare in modern surgery. Even when bruised plant debris, dried specimens that had been preserved for more than one hundred years, crushed fossils and finally powdered minerals all showed the same irregular vibrations his faith was not shaken. Smaller particles gave stronger movements because, as he believed, he was now isolating and observing the ultimate simple 'molecules' whose motion was a sure sign of their intrinsic life.

> These motions were such as to satisfy me, after frequently repeated observations, that they arose neither from the currents in the fluid, nor from its gradual evaporation, but belonged to the particle itself. . .The very unexpected fact of seeming vitality retained by these minute particles so long after the death of the plant would not perhaps have materially lessened my confidence, but that I observed on bruising the seeds that I greatly increased the number of moving particles. Reflecting on all the facts with which I had now become acquainted, I was disposed to believe that the minute spherical particles were in reality the supposed constituent or elementary Molecules of organic bodies.[1]

Nearly forty years were to pass before another agent for these movements could be suggested and this would transform the whole act of observation.

Nowadays this experiment is routinely demonstrated to every student in the school, although never in a raw and unprepared fashion for fear of an explanation similar to that of Brown but in an updated form – 'Are those the lung cancer germs in cigarette smoke?' Watching the demonstration may take no more than three or four minutes, but the preparation could well last a full hour. Only if this discussion with the class can successfully scale down the known and felt experience of all children to the size of the molecular universe will the random motion be 'seen' as the result of a frenzy of violent buffeting. Now it is the invisible bouncing molecules which are the chief actors in the drama and the small visible specks of matter become no more than their lumbering targets. The introduction to kinetic theory has radically changed what is being observed and has converted what is seen into convincing evidence for an entirely different order of activity.

Brownian motion is notoriously difficult to interpret, but school science is full of examples of phenomena where the abstract concept needs equally careful preparation. To demonstrate the enormous force of atmospheric pressure it is commonplace to extract the air from a tin can so that it will collapse dramatically in front of the whole class. The 'crunch' is always effective, but its cause is not self-evident. To the medieval world this would have given sensational support for their famous maxim that 'Nature abhors a vacuum.' To the Greeks it would have provided quite unnecessary practical evidence for their verbal tautology that 'what is not cannot be' (a vacuum would be non-existence). A modern child from the age of the vacuum-cleaner is more likely to hail the slow crunching of the tin can with delighted cries of 'The pump is sucking it in!' Only after the appropriate scientific indoctrination can they watch the same event with appreciative awe and comment, 'like a giant squeezing it from all sides'.

**Educating the Eye**

The abstract and invisible nature of the physical agency is a fairly obvious difficulty in interpreting an experiment. The actual visual features of observation would appear to present less of a problem, but they also can be dependent on what is thought to be observed. The everyday seeing of familiar things is painted in the colours of belief. There are people we mistake for friends when glimpsed in a crowd and patterns on the wall that can change with our moods from the outline of a dragon to the silhouette of distant hills. In the unfamiliar context of a new scientific observation it is by no means enough to suppose that a clear demonstration is even perceived by the untaught children as we can so clearly see it. There is continual intervention of unsuspected judgement between looking and seeing.[2] The human eye is only one part of what we call 'visual perception' and rely upon so heavily in the teaching of science. What the brain finally sees is not just a retinal image reversed like a positive print from a photographic negative, it has been unconsciously interpreted to accord with experience, expectation and hypothesis:

> The eighteenth century empiricist philosophers regarded perception as a passive process, whereas now we think of it as an active building and testing of hypotheses, which change to some extent throughout life. This means that we cannot, as they did, believe in raw data of perception and suppose that perceptually given 'facts' are solid bricks for basing all knowledge. All perception is theory-laden.[3]

Two fields of perception spring most readily to mind in the context of school science. The first of these is the world of the microscope slide. The youngest pupils first use microscopes to examine the cellular structure of plants. The practical details of this operation are difficult enough and are considered in more depth in Chapter 8. Eventually the focused image of a moss leaf or the stained skin of onion comes into view, but our problems are not yet over. Both of these may seem to present delightfully clear examples of the cells composing plant tissue – certainly they capture the attention of the young, so it seems appropriate to say nothing beyond asking for a direct drawing of what is seen. Not only the dictates of the discovery method but also a reluctance to interfere between the beautiful natural object and its fresh young observer tend to inhibit didactic comment. This can lead to dismay and disappointingly muddled sketches. Without the analogy of bricks in a wall it is rarely possible for the eye of a child to interpret what is before him as a collection of cells. The strong outlines of the cell walls can just as well suggest threads in a woven cloth or the pattern of a lace. Only when the eye is told to 'see' cells do the lines become outline walls and the unit cells themselves bcome the subject of the perceived image.

There is satisfaction in visual understanding and a patternless, incomprehensible picture can be distressing as well as confusing. This is true of many forms of visual perception. See, for example, the famous illustration where the eye wanders unhappily from one side to the other trying in vain to resolve the unresolvable. In the case of science teaching, where visual analysis and understanding are the paramount aims of all observation, help in 'seeing' the objects presented is often quite vital. Sometimes, for example, it proves necessary to present the drawing before the microscope slide in order to make perception possible, whereas educational discovery theory would insist on the opposite order.

Another obvious example of perceptual difficulty is the seeing of optical images. Many children assume that the view they see framed in a mirror is similar to that in a picture and they are led by this belief to

Figure 5.1

'see' the images painted, as it were, upon the surface of the mirror. There can be great difficulty for them in interpreting the visual evidence in such a way that the reflections of objects, which they so clearly observe, really lie behind the mirror where their 'common sense' insists that there can be nothing of the kind. They are taught to locate these images by placing a pointer behind the mirror and adjusting its position by the method of no-parallax until it appears to be coincident with the image, but this is no more than a procedural recipe. If they could trust the naked 'fact' of visual evidence, then locating the image would involve no more difficulty than placing a pencil at a given spot on a sheet of paper or picking a flower. Seeing a real image formed well out in front of a concave mirror is yet more difficult. The teacher may place a pointer in exactly the right position and claim that the illuminated image is clearly to be seen adhering to the pointer but the child is still blind to the whole effect. His brain insists that, since the mirror is the reflecting agent, the image must be formed upon its shiny surface and so he 'sees' it there. Neither by persuasion nor by slow degrees, but with a sudden visual click at a moment when the obstinate reasoning process can relax, the whole phenomenon changes over and the image appears at last in the indicated place. It is remarkably similar to the sudden inversion of a trick picture which can be seen in one of two different ways. Only the relaxation of insistence, not a deliberate effort, can cause the interpretation of perceptual evidence to change.

Figure 5.2

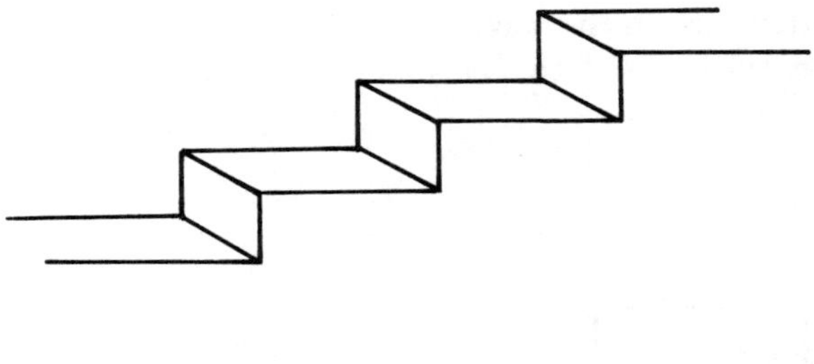

**Casting the Experimental Net**

There is a third way in which lack of conceptual preparation can ruin the effect of a school experiment, by influencing the order in which it is carried out and observed. Some phenomena produce slight or sudden changes which can easily be missed by watching for large or steady effects or concentrating the attention on some other part of the apparatus. It is relatively easy to prevent this waste of effort by some careful discussion and instruction before the experiment is begun. This might appear to make the lesson boringly didactic for the pupils but, even for these more practical hazards, it is not so much the details of activities that are required as an incentive to imagine the concepts and mechanisms underlying the experiment.

The most famous example of this kind of conceptual trap is the experiment performed in 1825 by Michael Faraday which *failed* to discover electromagnetic induction. He had laid out two circuits side by side to look for current generated in the secondary circuit by induction from that flowing in the primary. It is often said that it was the mere chance that he happened to close the primary circuit before closing the secondary one that delayed the moment of discovery for six more years. However, a reading of his diary shows that this first failure was produced by a conceptual fault in imagination and not by a mere practical chance. At this time Faraday had been influenced by Ampere's theory that two electrical fluids flowed in opposite directions around the circuit alternately combining and separating as the charged fluid particles passed each other on their different routes. Simple electrostatic induction should then have produced the 'mirror image' of this steady current flow in a neighbouring circuit. It would have been an undeserved stroke of luck for him to have captured electromagnetic induction in such a distorted conceptual net. By 1831 Faraday had a new theory that the starting or stopping of a current should generate a momentary ripple in the surrounding medium. This time he set up his experiment according to his new conceptual image so that it had make and break facilities in the battery circuit. His 'wave of electricity' was then duly discovered, as expected, by a sudden oscillation of the galvanometer needle.

Figure 5.3

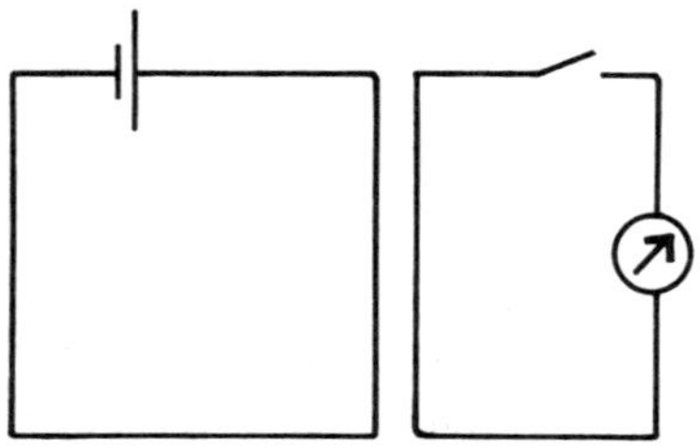

Figure 5.4

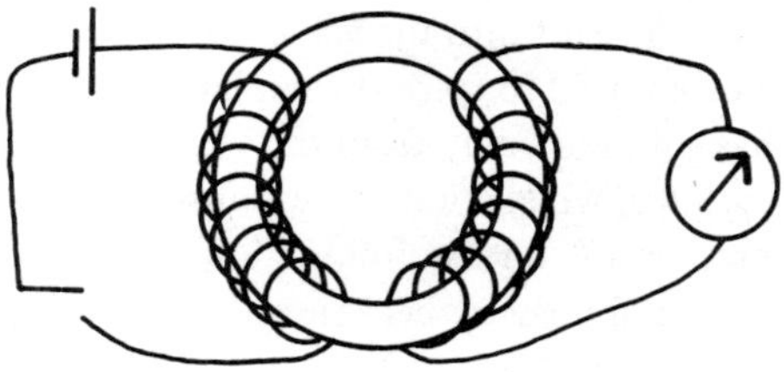

This very experiment, or one like it, needs to be performed in school. Can we now learn from Faraday's experience and, without jettisoning the discovery ideal by talking the whole experiment through beforehand, so order the demonstration that the required effect is perceived? We could, for example, take a large solenoid with a magnet inside it, show that it produces no current in the coil and then casually remove the magnet while the children are still watching the demonstration galvanometer. This certainly produces surprise and interest, but it also suffers from the same lack of conceptual preparation and is followed by a bemused reaction. Even if the pupils follow up the demonstration with experiments of their own to find out what factors affect the magnitude of the current induced, a sense of bewilderment still continues. Understanding a scientific experiment requires deeper interpretation which mere visual perception alone cannot produce.

**Learning through Association**

Enough has been said about the pitfalls and difficulties of experiments without conceptual understanding, but how can guidance be offered? Young children are rich in imagination, but the new scientific abstractions prove dry and difficult for them in the language of textbooks. Indeed no prescribed form of words will suffice. We need now to call upon some of the deepest magic in the whole teaching process, to reach out gently into the minds and experiences of the children, touching a thought here, a feeling there, jogging a few half-forgotten memories, until each one can begin to picture for himself an understandable and personal image. The practised teacher invents and tests out new methods of explanation as soon as his empathic 'feel' for the class assures him that the point has not yet been assimilated. The methods he uses to stimulate imagination are probably purely pragmatic, but studies in the fields of child psychology, philosophy and even physiology can give partial insight into the value of the mimicry and analogy which are often used to spice the explanation.

The term 'mental imagery' has suffered badly from the vicissitudes of psychological fashion. In the 1920s, when stimulus-response theories of learning were first being erected on the uneasy foundation of the rat-in-maze kind of experiment, mental imagery became an unacceptably romantic concept. An animal driven by hunger or electric shocks learns in an essentially blind and unthoughtful way: the form of the experiment insists that it shall be so. This is hardly likely to be applicable to the modern, liberal classroom! Curiously enough, it was other animal experiments, those of Köhler with his apes, which led to the rehabilitation of the mental image through the recognition of insight into a problem situation. No doubt the man in the street and the teacher of children had never relinquished the idea.

Köhler was a friend and follower of Wertheimer, the founder of the Gestalt theory of learning. In place of the older associative theory, with its tidy bundles of past sensations forming a conglomerate concept, the new psychologists began to emphasise the importance of organisational insight. This creates a system of imaginative groupings of learned items, concepts which are more than the mere sum of the items, and this in turn transforms the processes of perception, of committing to memory, and of recall.

Modern educational research has tended to examine the rather trivial effects of mnemonics or visual tricks for the recall of arbitrary learning sets. It is fairly easy to design tests along these lines and achieve quantitative results showing that such organisation of material does indeed increase the effectiveness of short-term memorisation by a factor between 1.5 and 3.[4] These internal devices for the pairing of unrelated items on a list may seem contrived, but in the field of science the creation of a greater meaningful structure will not only aid the process of learning a list of facts, it will supersede it forming an umbrella concept. Should this be a dynamic insight, it may even form the basis of a complete theoretical model. Clearly the problem is at the very heart of science teaching.

Studies on the child's capacity to learn language suggest that there may be some innate structure of the mind to which all language corresponds.[5] Such formidable linguistic or semantic organisations might also be used to make learning simpler. However, research seems to show that pictorial, visual structuring of information is more effective than verbal schemes. In practice, a verbal clue may well suggest a visual image so that these two factors, as always, are hard to separate. It may even be that their close correlation is itself innate due to co-ordinated or closely neighbouring regions in the brain.[6]

**Remembered Images Re-enacted**

What the science teacher needs is not so much a hypothesis of mental structure as a theory of the mechanism of mental imagery, if such there be, which will suggest a way to arouse dormant images. The outline of such a theory is to be found in the works of both Köhler and Piaget. Köhler was attempting to probe the brain to gather clues on the organisation which allowed learned principles embedded in a vague memory to be reconstructed to illuminate some new problem. He managed to detect slowly changing electrical potentials on the surface of the brain during the awakening of half-forgotten images which were similar, in some respects, to the electrical patterns produced by the original visual stimuli. Memory, he deduced, mirrors visual perception by 'dynamically equivalent' neural patterns.[7] With such a close connection between the remembered image and the act itself we have here the first clue to a method for reawakening the kinds of imagery we may need when teaching children.

Piaget's approach to learning was through the various stages of development that he observed in children. He saw the first two years of life as sensorimotor growth when the subjective experience of the infant, as he moves and senses the novel objects of his surroundings, forms the sole intellectual activity. His own movements and the reflexes of pain and balance are more vital and vivid to him than is the objective existence of the actual sharp points and moving objects. All the rest of the development of the child rests on this foundation. Slowly the older infant and the young child begin to investigate their environment more actively, forming their own methods of understanding it – egocentric and animistic at first, but always related to their sensory experience of objects.

> By virtue of the very fact that the child cannot translate his observations into a system of verbal judgements and reflexive concepts but can simply register them by means of sensorimotor schemata, that is by outlining possible actions, there can be no

question of attributing to him the capacity of arriving at pure proofs or judgements.[8]

When Piaget's work led him to examine the mental imagery of the older child he adopted a similar picture of 'internalised activity' or the imitation of it.[9] Like Köhler, he used neural evidence to demonstrate the sensory reactivation caused by the appropriate imagery. (We even produce almost identical electroencephalograph traces by conjuring up the mental image of flexing the hand as by the act itself.[10]) These sensorimotor images of remembered personal activity can prove of exceptional value in learning science when, as is so often the case, there is a mechanism to be understood to which they can be linked.

**Piagetian Theory**

Most of Piaget's experiments on images were related to mere visual perception of static objects which, for reasons that are mentioned later, prove less important to the comprehension of experiments in school science. However, he described a fascinating series of experiments he had carried out with young children watching the progress of a somersaulting cardboard tube which brought out very clearly the varying degrees of ease with which mental images can be stored and recalled. The tube was placed on a table with its end protruding. It was then simply pushed down at this end so that it turned over in the air and landed on the ground in the reverse position. After the demonstration the watching children, aged between 4 and 8, were set three tasks:

(a) to draw the stages of the tube's path;
(b) to reproduce the movement of the tube in slow motion by handling it; and
(c) to describe the motion of the tube verbally.

Now if the image in the child's mind were simply a slowly fading trace of visual perception, the drawings in (a) should have been easy to reproduce, but in fact this proved to be the most difficult of all the tests. It was the manual imitation of the tube's motion in (b) which scored the greatest number of successes, lending considerable support for the existence of an internalised, non-verbal, motor image.

Piaget's theory of child development has had a large influence on

Figure 5.5: The Somersaulting Tube!

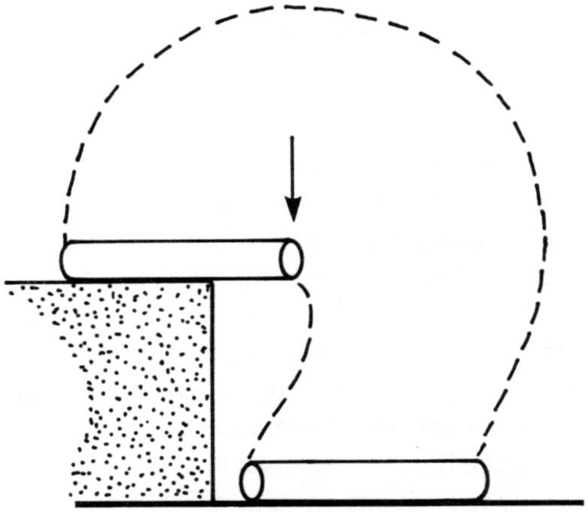

American primary school teaching. In particular, their SCIS (School Curriculum in Integrated Science)[11] is so steeped in his philosophy that almost every experiment and every response is labelled by the appropriate growth stage in the teachers' manual. However, his ideas on mental imagery have never, to the best of my knowledge, been applied to the field of secondary school science. At this age the pupils are beyond the stage of those involved in the test with the twirling tube and are facing another hurdle. In Piagetian phraseology the secondary school is attempting to explain science in 'concrete operational' terms of possible manipulation and reversal, and then to wean their pupils to the final level of formal reasoning.[12] By this stage the concept itself would be so well understood that it could be used to predict and explain further scientific phenomena. Despite this ultimate aim, the earlier stage of cognitive development must not be neglected both because it is deeply familiar to the teenage pupil and because the change-over to formal thinking is uncertain in its time of onset, appears to suffer periodic eclipses and may, unhappily, never be reached at all by many of the children. In particular, when we are introducing totally new scientific ideas this power of formal reasoning, even if it has already begun to develop in other fields, is not the faculty we require. It has nothing yet to build upon. Dawning comprehension needs an analogy to a familiar effect whose felt sensorimotor aspect has already been thoroughly assimilated in early childhood. Only then can thought progress beyond the activity to the abstract concept and so to its use in a reasoning situation.

## The Feeling Experiment

The experiment itself presents the visual evidence and it is easy to suppose that this stimulus will trigger some effective memory and so establish the required conceptual understanding. This belief is enshrined in the precept 'Seeing is believing', but a glance around the puzzled faces behind the laboratory benches shows, all too often, that this is not the case. The experiments on Brownian motion, collapse under atmospheric pressure and electromagnetic induction already mentioned, as well as a host of others, illustrate the breadth of uncertain territory between observation and understanding. Seeing alone is not enough. The modern child is shown daily marvels and visual tricks, but his scepticism has deeper roots than mere sophistication. The liaison between eye and brain is a kind of collusion with such a ready interpretation and even distortion of the retinal image that we all instinctively feel a certain doubt about what we see.[13] Our eyes too often deceive us: 'Is it real?' we ask, and reach out a hand to touch it.

Any experimental work concerned with movement or some working mechanism will need to impinge more directly on the faculties of the participants than does a visual spectacle, however dramatic. The series of experiments routinely carried out on force, acceleration and velocity are a case in point. This work was enthusiastically presented in the earliest of the Nuffield projects and firmly based on their slogan that only by 'doing' will understanding follow. In practice this part of their course is probably the least effective as it stands (see Chapter 8), difficult to carry out, long-winded in calculation, often inaccurate in conclusion and, worst of all, obstinately visual in its impact. It is the trolley, not the child that experiences the force and the acceleration. After all the calculations have been made and the graphs plotted, many children remain unconvinced. If their relations with their teacher are sufficiently confident and easy, and if they are keen enough to ask the same incoherent questions again and again, it will become clear that the results remain unassimilable. This is, perhaps, the moment to borrow a large trolley from the kitchen and wheel a child across the laboratory with acceleration rough enough for him to feel, in his own body, the jolting force involved. The others will watch and remember their own familiar reactions to starting and stopping buses. Once this personal motor image is awakened it can be transferred, as it were, to the insensible trolley of the experiment.

The same motor image can be used as an analogy. The experimental demonstration of electromagnetic induction, for example, which is so hard for the children to comprehend simply from the visual effect on the galvanometer, begs for a similar explanation. Electrons, of course, do not behave like the human body, but a change in magnetic field can be compared with a change in momentum as if the free electrons in a metal 'feel' this and are jerked into movement like the passengers in an accelerating car. Imaginatively presented, this kind of analogy to the 'internalised imitation' of a sensorimotor image can be very helpful to many children. Acceleration is often sensed with uncomfortable vividness in the pit of the stomach and even an older pupil, beginning to study the mathematics of simple harmonic oscillations, can be helped in an imaginative way by allowing this delicate organ to follow in sympathy with the slow vibrations of a weight on the end of a spring!

## Body Memory of Movement

Reawakening these motor images can be a delicate task. Not only are they laid down in early childhood, but they are almost always unverbalised. Their value lies in their intensity and credibility related, as they so often are, to life-giving body reflexes towards pain, touch or balance. We know that other, more complex patterns of movement, like playing the piano or riding a bicycle, remain in the brain as 'motor memories' which are notoriously hard to credit until the activity is actually begun. We could swear we have forgotten how to play a piece until our fingers are actually on the keyboard, or to cycle until our body reacts to the movement of the bicycle. No conscious act of memory is as effective for awakening this motor understanding as is its re-enactment.

What knowledge we have of the siting of different kinds of memories within the brain also shows that there is no simple system of pigeon-holes. Evidence from the examination of patients with serious brain lesions has demonstrated that a process which is learnt in one part of the brain can move about during the development of the child and be stored in quite another part.[14] As far as motor memories are concerned, Luria's work suggests that there are at least two different areas involved, corresponding to the more infantile reactions and the later learned skills.

> Essentially the kinetic organisation of movements is performed by completely different brain systems: by the basal ganglia in the early stages of phylogenesis (the stages of elementary 'motor synergism') and by the premotor area of the cortex in the late stages of the formation of motor skills.[15]

If we add to this Luria's evidence that the localisation of the higher mental processes is never static, but moves about within the brain, we can appreciate the difficulty of stimulating the images that we need by any but the most direct of methods, aimed at the autonomic regions of the ganglia themselves.

All science, not just that which is taught at school, is rich in imagery, suggesting that the right way to understand it is by an inventive use of the imagination. This should be no great problem for children so long as we get the analogy right. A surprising number of these concepts can be acted out instead of just being talked through so that these precious motor reactions are involved. Some, like compression waves, attraction and repulsion, even convection currents, can be mimed by the children themselves, while others, like replacement of metal cations and resonance can be suggested by a word or gesture on the part of the teacher. For the pupils it may seem no more than a light-hearted interlude, but there is no mistaking the effect it can have. There is plenty of scope here for the individual approach by a lively and uninhibited teacher!

**Can We Touch It?**

Closely related to motor stimulation is the sense of touch. In the earliest stage of infant learning this is a very vivid sensation – many children of primary school age still react to the touch of velvet with an actual shudder. The speed at which a blind adult 'reads' Braille shows the sophistication that is possible, but in sighted adults this sense is so little used that it tends to atrophy. Interesting studies[16] which have compared the use of the senses of sight and touch by workers selecting wood for the fineness of its finish have shown that the inexperienced worker not only gains no advantage by the exercise of touch as well as sight, he actually loses by the addition. Other studies[17] on the discrimination of length have also demonstrated the same curious result; there was a *greater* error when both senses were used together. The authors of this research concluded that 'in novel or ambiguous situations there is no coordination between visual and tactile impressions.' There seems to be some general intuitive understanding of this odd physiological effect; feeling works better without looking. Even at college or sixth-form level a group of crystallography students urged to examine the bonding directions in a large crystal model will often close their eyes spontaneously as they feel the direction of the 'spokes' between the 'atoms'. Whenever the school situation calls for the use of the tactile sense this effect should be borne in mind. Mass or inertia, for example, is a very hard concept to grasp and one of the best ways to reconstruct the appropriate image is to place a kilogram weight on a trayful of ball-bearings and have a child feel the ease with which it can be accelerated and stopped when compared with two such weights stacked on top of each other. Once again the way to profit from this tactile experience is to close the eyes. A gentle suggestion from the teacher is usually enough even for a timid pupil because it accords with some unspoken experience of childhood.

There are splendid cases where the sense of touch, like the motor sense, can of itself evoke the conceptual image when purely visual evidence remains obstinately unconvincing. Pressure within liquids is a static tactile experience. A junior class presented with the hackneyed experiment involving a tall container full of water and pierced at several levels can easily observe the trajectories of the different jets produced. By projecting their shadows on to a screen they can be traced or drawn, but this rather beautiful visual effect does not immediately construct a notion of fluid pressure. With these young children it is far more effective to give them a polythene bag to put over one hand and then encourage them to plunge it into a bucket of water to feel the squeeze at different depths. Most of them will have already waded into rivers wearing Wellington boots or have washed up the dishes in rubber gloves. The inarticulate idea of pressure is lying somewhere deep in the mind ready to be used as soon as its tactile memory is reawakened and linked to a new conceptual name.

## New Motor Experiences

These long-familiar sensorimotor images are valuable and strong in spite of the fact that they are unverbalised, indeed perhaps just because of it. In later life we experience relatively few new motor effects and assimilate even fewer images. When learning to drive a car we force ourselves to accelerate round bends instead of instinctively braking, and

Figure 5.6: What *is* Pressure?

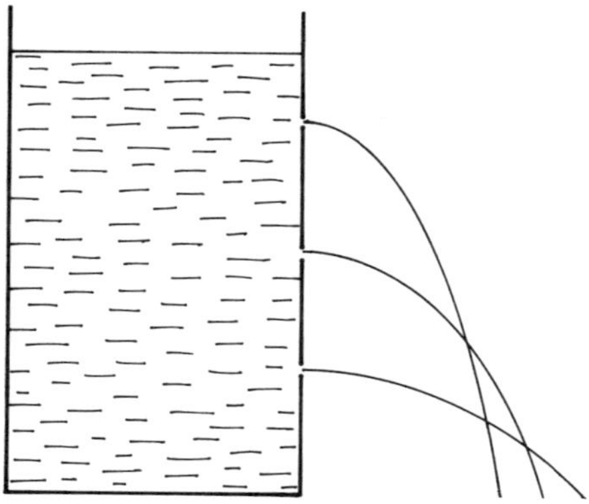

can only explain rather lamely that it seems to help the car to 'hold the road'. There is no immediate body-understanding of the reduction in force. It is fun to learn about moments of inertia with the help of a revolving chair, and to feel the increase of angular velocity as the arms and legs are pulled in. This lends credibility to the topic, but the effect is found to be curious rather than familiar. Only if the class includes ice-skaters or ballet dancers will it jog any recollection. New motor feelings can well prove distressing as well as incomprehensible, for the adult body is a conservative creature. The first astronauts, for example, had been extensively trained in short spells of weightlessness and then, on their journey out to the moon, spent three whole days without appreciable gravity. On arrival at the moon Buzz Aldrin's reaction was one of sensorimotor and emotional relief: 'The moon was a very natural and pleasant place to work. It was less lonesome than zero-gravity because you always had a distinct feeling of being *somewhere*.'[18] Science teaching needs to be well spiced with allusions to practical effects to keep strong links between science and everyday realities, but these more sophisticated motor illustrations form a different teaching aid from the motor *memories* entrenched in the mind since infancy.

### The Roots of Understanding

As the student progresses through the science curriculum he will encounter ever stranger theories at a further remove from daily experience, but still learning science must not relapse into the Red Queen's practice of believing six impossible things before breakfast! Whether it is a drifting electron gas within a hard metal, discrete quanta of energy or the amino acid matching of RNA – every new wonder is an image to be examined and assimilated by an imagination already trained on simpler fare or else the whole edifice of theory will become unsound. The most abstract of theories have grown out of our everyday images, and these are the essential common ground which permit us to teach, compare and communicate ideas. 'At the outset, the consensual foundations of science rest firmly enough on the universal cognition of real world of everyday things. This is the starting point of all scientific education'.[19] For school experiments, too, we need not only the hardware and the know-how to use it, but also the wealth of chaotic images which our children possess for providing the material to shape essential concepts. The attempt is always to help the pupils to integrate their experiments with a common core of experience in order to clothe it with a *personal* reality.[20]

The way in which we set out to awaken these personal images is varied. The examples already presented have ranged from the easy gesture of a teacher to a whole class activity[21] in the hope that these will trigger the child's own memory and understanding. Occasionally a pupil will ask directly for help in understanding a concept, but more often it is left to the teacher to scan the rows of faces for the blankness that shows incomprehension as school science becomes alienated and 'separated from their non-school knowledge of the natural world'.[22] Probably the best method of locating these trouble spots is a short piece of imaginative writing. The more usual homework – a description of an experiment or the working out of a problem – is sadly dry for young children; it also contributes to the alienation[23] of school science from general knowing, as well as wasting precious childhood qualities through lack of use. The 12-year-old who confides hesitantly, 'I imagine it like this . . .' is treading the real road of scientific research.

## Notes

1. Robert Brown, *On the General Existence of Active Molecules in Organic and Inorganic Bodies* (1828).
2. See Richard Gregory, *Eye and Brain* (Weidenfeld and Nicolson, London, 1972) for plentiful illustrations of this.
3. Richard Gregory, *The Intelligent Eye* (Weidenfeld and Nicolson, London,

1970).
4. A. Paivio, *Imagery in Verbal Processes* (Holt, Rinehart and Winston, New York, 1972).
5. N. Chomsky, *Language and Mind* (Harcourt Brace and World, New York, 1972).
6. J.M. Ziman, *Reliable Knowledge* (Cambridge University Press, Cambridge, 1979).
7. W. Köhler and R. Held, 'The Cortical Correlate of Pattern Vision', *Science,* 110 (1969).
8. J. Piaget, *The Child's Construction of Reality* (Routledge and Kegan Paul, London, 1955).
9. Jean Piaget and Barbel Inhelder, *Mental Imagery in the Child* (Routledge and Kegan Paul, London, 1971).
10. H. Gastaut 1954 (Clin. Neurophysiol.).
11. Not to be confused with the British SCISP for secondary schools.
12. R. Fuller, P. Karplus and A. Lawson, 'Can Physics Develop Reasoning?', *Physics Today,* 30, 2 (1977).
13. Consider the well-known experiment with a volunteer in inverting spectacles who actually learnt to beat the glasses by 'seeing' the wrong way up!
14. I. Luria, *The Working Brain* (Penguin, Harmondsworth, 1973).
15. Ibid.
16. R. Harper, *Human Senses in Action* (Churchill Livingstone, Edinburgh, 1972).
17. Kelvin and Mulik, 'Discrimination of Length by Sight and Touch', *Quarterly Journal of Psychology* (1958).
18. We Came in Peace, NASA Publishing, 1971.
19. Ziman, *Reliable Knowledge,* CUP, 1979.
20. 'This conception of knowledge as personal knowing departs in two closely related respects from the ideal of strictly justifiable knowledge. It accredits man's capacity to acquire knowledge even though he cannot specify the grounds of his knowing, and it accepts the fact that his knowing is exercised within an accidentally given framework. These are correlated within the effort of integration which achieves knowing. . .To this extent knowing is an indwelling.' (M. Polanyi, *Knowing and Being* (Routledge and Kegan Paul, London, 1969).)
21. *Nuffield Combined Science Activities,* p. 71.
22. Michael Young, 'The Schooling of Science' in *Explorations in the Politics of School Knowledge* (Driffield Studies in Education Ltd, 1976).
23. See also 'thought experiment' in Chapter 9.

# 6 BEHAVIOUR AND TEACHING STRATEGIES

## What is Discipline?

The beginning of a science lesson is a particularly difficult and important phase, and this is not just for reasons of pure learning logistics. The reality of teaching a class of unpredictable and lively children can be rough and even frightening, at least for a novice teacher. A sympathy and love for the freshness of children and a desire to fire them with our own delight in science are marvellous assets, but to ignore or under-rate the ebullience, sullenness or sheer contrariness of children does no service to the cause of teaching and indeed robs it of both elation and despair. As these first lessons begin, we shall need to use every weapon at our command to encourage, instruct, stimulate and curb, not only to maintain a feasible level of good behaviour but, interwoven with this, to teach the principles of science. Each objective can help the other. Discipline for its own sake is sterile, and learning is impossible without it – but the ways of effective learning can be a discipline in themselves.

The paramount requirements of the learning environment are *attention* and *interest*, the only substance to that harsh archaic word 'discipline'. Perhaps there was a time when the aura of fear, painfully sharpened by instant corporal punishment and religiously instilled respect for age regardless of merit, could generate an atmosphere of terrorised attention whatever the subject-matter being taught. Such an achievement runs counter to our notions of happy learning and free curiosity; let us note its parting with satisfaction, not regret. The modern school has very few effective punishments, 'sanctions' or 'negative inducements' as the current jargon has it, with which to coerce its pupils. Reporting a malefactor to the Head or some other hierarchical figure brings little added authority to the teacher, and extra work after school is hard to enforce but can help to enlighten a child and bring a little closer contact between him and his teacher. In the long run attention can only be achieved positively via our other objective, interest, and this we can quite profitably attempt to fashion.

It is often imagined that an old-fashioned kind of discipline can still be obtained by the mythical 'born teacher' through his innate qualities of charismatic authority and love of children. Such beautiful attributes would be hard, even for a saint, to sustain daily through every crisis. For the rest of us there is just the confidence that a framework of well-organised activity provides, together with the tools for matching it to the behaviour characteristics of the class.

**Learning Methods and Behaviour**

This chapter will rely no more on the theories of child psychology than on the hypothetical character of the ideal teacher. Instead we will keep our gaze firmly fixed on the possibilities of the teaching programme both for discipline and for the learning of science. There are always two parallel avenues to explore:

(a) the best learning steps to be rehearsed for ease of understanding;
(b) the ways in which these steps can be applied to match, in a remedial sense, the mood and behaviour of the class.

The conduct of the lesson is no less free of ideological argument than is its content, indeed this is the very kernel of the conflict between the 'discovery' and the 'expository' methods. The rival approaches are among the alternatives possible, but judgement must rest not upon theory but upon the practical criteria stated above. Some element of discovery was always present, and even in the most traditional form of lesson the numerical value of a coefficient or the identification of an acid radical was expected during experiment. It was the start to these lessons which was invariably expository. The topic was stated, underlined on the blackboard, and taken down in notes at dictation speed before the method of experiment was carefully demonstrated for the pupils to emulate. It is still a viable method although, in its extreme form, any calling out by the pupils, even helpful suggestions, could be a punishable offence. Its advantages, as we shall see, are primarily those of a memory reinforcement and an activity depressant when used in the appropriate circumstances. Its disadvantages are very clear; it rates low for the arousal of interest and personal motivation, and it makes the teacher's task of adjusting his tempo to the rate of comprehension of the class doubly difficult in the face of such an inhibition of feedback.

Of course this method is anathema to exponents of the discovery ideal. They see such a lesson as an artificial, encapsulated, teacher-given lecture without any inducement for the children to think out problems on their own and find for themselves the delights of 'discovery'. Their favourite methods too have a place in the craft of teaching, but as tools to shape attentive behaviour in an unruly class they can prove a hideous pitfall for too dogmatic a teacher.

Many of the suggestions to be made may be of equal application to the teaching of other school subjects, but science lessons do have some special advantages as well as special hazards. The subject-matter is often closer to the lives and interests of the children than, for example, history or foreign languages. The experimental work in particular is welcomed by the pupils as an exciting break from continual sitting and writing. On the debit side, however, it needs little stressing that violent and undisciplined behaviour is extremely dangerous in a laboratory and, less obviously but equally serious, bored inattention can also lead to accidents. Neither can be tolerated during practical work. However strongly urged by some well-intentioned educational psychologies, no apparatus should even be handled while such hazards seem at all possible. Inquiry or guided discovery learning imply a measure of teacher control no less real for being less apparent than the deathly quiet of traditional discipline.

## The Varieties of Behaviour

The spectrum of behaviour in children is more one of activity than of pure villainy. They can be more energetic and volatile than adults, but they also descend far more readily into total apathy. As we attempt to shape a lesson, a variety of different modes will spring to mind which can be used not just for variety's own sake but as instruments to adjust the level of activity. Too little will produce half-hearted work or none at all, while too much, if unlinked to the problems being studied, may lead to that wild aggressive behaviour which is the nightmare of teaching. Shouting, fighting and destruction would be pathological in an adult – in children they lie quite naturally at one end of the normal range of behaviour. This frays the spirit and makes teaching near impossible, but we should study to contrive a charity of spirit (though not toleration) towards this unattractive aggression of the young.

> human children need all the aggressive potential they can muster to protect and assert their developing individuality.[1]
>
> At origin, aggression is almost synonymous with activity.[2]

All we can win from the words of such eminent child psychologists is the reassurance, if such it be, that this is a natural, even a healthy characteristic. Unfortunately it is no part of their task to offer advice on how to react when twenty or thirty noisy youngsters are displaying this healthy potentiality simultaneously! We can do no better than to choose our teaching methods to galvanise the lethargic or quieten the hyperactive so that the extremes of sullen boredom or near violence are moderated towards a useful level of activity.

There are three main types of approach to the start of a lesson which we could label 'the demonstration', 'the inquiry' and 'the recapitulation'. Individual teachers may tend to be wedded by continual practice and personal preference to just one of these, but all three deserve consideration both for their learning potential and for their influence on class behaviour.

## Demonstrations for Teaching

A demonstration at the start of a lesson sounds as conventional and old-fashioned as the dunce's cap, but this is misleading, as it can serve many different purposes. In the simplest mode it gathers the pupils together around the teacher to look at a single exhibit which is more complicated than those they can manage on their own. A dissected animal has an emotional impact on young children. They find it both fascinating and horrifying, to the point where these intense feelings stand between them and the information they could be gleaning. In this situation the teacher needs to be close to his class, talking carefully with reassuring, familiar words, pointing to the intricacies of the organs displayed and showing their connections and functions. Questions about killing the animal and expressions of disgust can be pacified more easily for a group of children whose physical closeness makes communication by word, tone and gesture more intimate and immediate.

There are big demonstration experiments that the children should be shown. An 'electron beam' apparatus, a model steam engine or a Liebig condenser are too intricate or expensive for the children to use on their own. We will set it up for them but even if it is still visible at the back of the class the length of the laboratory can easily produce a feeling of detachment. Once again we need to have the children close by, not only for psychological reasons, but also to observe the details of the apparatus and to become involved as one or two volunteers lend a hand with the work. In young classes the invitation to gather round is received eagerly, to the point of clambering on the benches and almost throwing themselves into the apparatus! An older class can be irritatingly lethargic in its response, but physical movement may of itself break the immobility of thought and generate a modicum of participation in even the most blasé of pubescent pupils.

Demonstrations serve another purpose; they teach the use of new apparatus and new procedures. The facile worksheet may appear to give foolproof instructions, but many of our children are not nearly as literate as we could wish, and what may seem unambiguous to the writer turns out to be riddled with possible misunderstandings – even normally intelligent adults can be reduced to frenzy by following instructions for assembling a simple kitchen gadget! All new equipment will need a technical introduction, from the method for connecting a circuit to the use of a blow pipe and charcoal block. We don't want to steal the surprise of the experimental result, but we do need to be sure that it will be conducted in an appropriate way. A demonstration of this kind whets the appetite for individual work, acts as an incentive for accuracy and ensures a higher success rate than do the written instructions alone. The prospect of being about to do the experiment on their own is enough to produce careful attention until eagerness makes them restive to begin. The discipline problem is not usually one of tampering with the demonstration; scientific apparatus looks more imposing than it is and their chance to use it is in the offing. However, children packed together are not so easily watched and tempting small items of apparatus (such as magnets) should not be left close at hand while the teacher's attention will, necessarily, be on the conduct of the demonstration.

## Demonstrations for Surprise

Finally, a demonstration can be used not so much to teach as to startle. Instead of a sober pedagogic experiment the children are presented with a trick – in front of their very eyes the impossible seems to be happening! Teachers and lecturers have often produced bright lights or loud bangs, all children love them, but the aim of this approach is not so trivially entertaining. The American educationalist William Ramsey,[3] who writes about this method, recommends specially that it be used to 'offend the intuition' of the young onlookers, which will excite them into interest and challenge them to advance their own explanations for the trick. From this they are then drawn to design their own investigations.

Examples will make this clearer. Ramsey's own suggestions include pouring some indicator solution into three apparently empty beakers, whereupon it mysteriously changes colour, rousing wonder and curiosity. The children will immediately suspect that the beakers were not so innocently empty and offer to examine them. The teacher's job is to channel this curiosity into a stretch of useful work on pH values and neutralisation. Another invention of his involves dropping ice cubes into beakers, both apparently containing water, only to find that they sink in one case and float in the other. Again the children recover from their initial surprise with the canny suggestion that one of the liquids is not what it seems to be! They might even offer to examine its density and set about listing the apparatus that they will need. Such, at least, is the hopeful philosophy of the method.

Some teachers object strongly to this approach on pseudo-ethical grounds. Science, they maintain, like their own teaching image, should be seen to be truthful and serious, and they decline to be cast as mere conjurors. They would like to demonstrate always, by word and by deed, that experimental results are our most reliable source of evidence and that their use for deceit, even as an educational ploy, is inadmissible and damaging to the prestige of science itself.

Of course each teacher is entitled to employ his own style of teaching, but it is not necessary to wield so heavy and purist an argument. This is most unlikely to be the regular format of lessons, both because of the difficulty of continual invention and surprise, and also because it omits a lot of the necessary steps in the learning process. Its sole objective is motivation for experiment when the learning prerequisites for this have not yet been met. Our pupils will need:

(a) a recall of related learning;
(b) an introduction to the new concepts (acidity, pH value, density);
(c) familiarity with the apparatus.

Without these, discussion can only be incoherent and experiment ineffective. However, it would clearly be absurd to produce such a deliberately outrageous beginning to a lesson only to calm down and lose all the excitement we have generated in order to instruct the class in the missing material. The order seems at fault. Contrary to appearances, this is not a starter at all, it can only be used in the middle of a run of lessons on some topic in which the preliminary work has already been done.

Under the right conditions and with a generous measure of showmanship this approach can have real behaviour value. Watching a demonstration with attention already requires a fair measure of quietness in a class, but the great value of spectacle lies in its power to arouse activity. For an average to lethargic class temperature such a method provides a tonic and a sense of fun. In more rowdy, inattentive conditions we shall seek another method of beginning the lesson.

**Inquiry by Worksheet**

In dubbing the second method of beginning a lesson with the general heading 'inquiry', a number of different approaches have been loosely grouped together. They have in common only some element of individual observation or practical work:

(a) discovery experiments;
(b) exploration of an exhibition;
(c) experience type experiments.

There is some pressure to begin with practical work. Many children, especially the younger ones, habitually hurl themselves into the laboratory with exuberant cries of 'What experiment are we doing today?' It is well to take a cue from this and have something up our sleeves either for the start of the lesson or for later on. As has been indicated in Chapter 2, pure, open-ended inquiry is too demanding and deflating to the spirit to be given any recommendation. Every experiment requires a careful preamble and a set of instructions. There is a strong inducement to begin the lesson with a brisk handing out of worksheets full of instructions so that the children can launch themselves straight away into their own guided investigations. A good worksheet may appear to cover every point but, from a teaching point of view, this efficiency is an illusion. Once more both the recapitulation of useful previous learning and the careful introduction of new concepts have been totally ignored. The hopeful insertion of a written item further down the worksheet such as 'When before have you met a similar idea?' or 'In Unit 5 you will remember making similar crystals' will be far too late and too mild to add any significance to the gathering rhythm of the schedule. Only too often these drilled experiments conclude, like an obstacle race, with a finishing post but no achievement.

The behaviour rating of these recipe-experiments is harder to sum up. Most children, including the boisterous ones, relish experiment and will forgo all the pleasures of gossip, argument and mischief for the sake of a chance to 'make things happen'. No threat of punishment is as effective an inducement to listen to instructions as the promise of an experiment; it is probably the teacher's finest carrot! Nevertheless the choice of such an experiment needs considerable care if it is to proceed in a reasonably serious fashion with so little preparation. Not only must its concepts be familiar or very simple, its procedures must also be clear, foolproof and conducive to a quiet, 'learn-it-yourself' atmosphere. Bitter experience has shown that not all curriculum planners guide us with sufficient care in the selection of these experiments. In the Nuffield physics course, for example, there is a notorious introduction to wave motion by the use of ripple tanks, which can involve plunging an energetic class of third-year pupils, just recharged by the long summer holiday, into a blacked-out laboratory full of unstable water-filled ripple tanks complete with throbbing stroboscopic lights. In the early days of the Nuffield scheme this gave rise to scenes of devastation in some laboratories, including that of one new young teacher whose wet and darkened pandemonium was abruptly ended by the entrance of an irate Headmaster who had been rudely drenched in his private sanctum on the floor below the physics laboratory! For the same aggressive age group the Scottish Integrated Science Course publishes a worksheet of individual experiments entitled 'Pairs of Forces'. This blandly begins with the instructions to face your partner, both standing on one foot on a trolley or roller-skate and *push*. 'What happens?' inquires the worksheet. . .broken noses?. . .general fisticuffs? It would be hard to imagine any class so docile as to profit from this beginning to a lesson.

## Inquiry by Exhibition

In primary schools it is commonplace to allow nature study to arise quite naturally out of the specimens that the children bring to school with the familiar cry of 'Look what *I've* found!' At secondary school stage this ready contribution and delightful curiosity flags just when we most need to stimulate careful observation. The curriculum becomes more directed, more specialised objects are required and the teacher, not the pupils, will have to provide the collection if a special teaching point is to be made. (Odd biological specimens from the lamented wounded bird to the flea-ridden remains of a run-over hedgehog still make occasional appearances.) Most of the Nuffield courses begin with an exhibition of materials and specimens put out for the children to look at, feel and handle. In accord with the full freedom of their philosophy, no directed activity is specified, the children just wander round the display. Balances and rulers are supplied, but their use is not pressed. This is an exhibition for acquaintance only. 'Science is rooted in observation, so our course may well start with a gathering together and extension of children's experience.'[4] Watching these young children flitting from one sight to the next, one is forced to conclude that neither serious observation nor experience is the likely outcome, but only a hopeful sharpening of curiosity and a predisposition for this new field of study.

Such exhibitions have no narrow teaching purpose, but there are a few definite points in a science course where a tighter collection is called for, for example classification of the animal kingdom or the construction of a 'key' for identifying members of a smaller group, trees, butterflies, breeds of dogs or laboratory glassware, and this is best carried out from a layout of specimens or photographs in the laboratory. A course on earth science will require boxes of assorted and labelled rocks and crystals, preferably free for handling, and an introduction to the idea of chemical elements also suggests an exhibition of examples.

Even in this mode we can generate more or less activity by varying the conduct of the exhibition. Pupils can be encouraged to walk around from one exhibit to another at their own speed, or they can be seated at their places observing their own sample collection. Some restless younger children welcome the opportunity to move, and live animals, in particular, are a favourite with this age group. On the other hand, a mere physical orbiting of the exhibition, given their brief powers of concentration, will not serve. Writing or drawing slows down the kaleidoscope of impressions to a learning rate, but this is not easily done while standing pressed into a tight knot round some popular exhibit. Directed changes of activity keep a class on its toes, fully involved and free from the ever-ready menace of boredom which pulls down very effective shutters on young powers of observation. Writing games can be played on the basis of an exhibition; key classification as a challenge by the teacher, or a variety of 'Kim's Game' for naming some chemical elements. Other forms of activity, such as drawing the shapes of crystals or scratching minerals to make a hardness scale, generate more involvement in the exhibition and are useful for older children. A lot of preparation may be needed for these lessons, but they usually run well –

with one proviso. The teacher will be fully involved; young children call out frequently, eager to point out some item they have just observed or to ask about an unfamiliar item which has baffled them. Inevitably this means that other children will not be supervised and some additional help, perhaps the presence of a watching technician, may be needed to keep an eye on the more precious exhibits and on those pupils who may want to test the strength of fragile items or simply splash each other with the water which is always so temptingly to hand in the laboratory!

## Inquiry by Experience

There is another kind of practical investigation that is, at heart, as free-ranging as inspecting an exhibition. We need to evoke and name for our pupils some already half-familiar concepts and this is best done by using the simple senses of touch and feeling (see Chapter 5). Take, for example, the concept of *force*. The old Newtonian definition of 'that which tends to change the state of motion of a body' is hopelessly abstract, full of logical holes and totally incomprehensible to our children. But we have no need to explain this word – they all understand it already. Pushing, pulling, straining and lifting are a part of living, so the teacher's task is not to teach but to organise activities that will allow our pupils to refine and quantify their experience. This is an essential part of their course of study. To leap this stage is to build a heavy tower on verbal sand when the rock foundation is already to hand. Radiation, acidity, speed, concentration, density, breathing and gravity are just a few of these concepts which we can reawaken by the appropriate practical experience. This certainly belongs somewhere in the opening stages of any new topic.

Most of these ideas could well be recaptured by a period of simple play, but we cannot expect self-conscious teenagers to respond seriously to such a degradingly infantile suggestion! For this reason we concoct busy pseudo-experiments.[5] For an introduction to forces we hand round spring balances already marked out in Newtons, and since the tension of the spring is so easy to feel, we start by having them hold this balance in the hand while they pull. They can measure the force that their little finger can exert, the force required to hold up a kilogram weight or to pull a stool across the floor. Only after this do we take the balance out of their hands and have them hang it up. Now they can feel the forces of attraction and repulsion between two magnets and, by placing one in the scale-pan beneath the spring balance, they can measure these forces too. Few of the numerical results are invariable or of any lasting interest, although our pupils are not to know this. For them there is the scientific activity of quantifying and reading scales; for the teacher, the secret objective is the feeling and consolidation of an abstract concept.

The same method can be applied to the study of heat radiation. We can confidently assume that every child has held out a hand towards the fire or has shielded himself from the heat of the sun, but even an adult would be hard put to define such radiation in words. At this stage it is not even useful. The children can heat the corner of a gauze red-hot (they like this) and then we ask them:

> Which part of your hand – back, palm or finger tips – is the most sensitive to this radiation?
> Measure, in cm, how far away from the gauze you can still feel the heat.
> Turn off the flame. Does the gauze still radiate heat after it has stopped glowing?
> How far away can you feel it now?
> Can you see heat radiation?
> Find out whether this radiation goes through wood or through glass, etc.

The continued pressure of direct questioning keeps the activity busy, which is important in a mischievous class that can become unruly with more open-ended discovery. Little introduction is required, our only teaching may be a familiar gesture to reassure the timid that their commonplace sensations are truly valuable in science and as reliable and universal as language itself. Each child feels for himself and writes his own report. The teacher's main task is to ensure that incendiary material (spare wooden splints or strips of paper) are not misused as he wanders around the groups encouraging those who need it and trying to infect those who rush on too blindly with a little more care and wonder.

## Recapitulation and Forgetting

Recapitulation does not sound a very stimulating beginning to a lesson, and certainly for sheer excitement it has much less to offer than either a demonstration or an inquiry. Nevertheless it can be handled in a number of imaginative ways: it has great value as a behaviour moderator and its learning effect is quite vital for a number of reasons.

In the first place, and most obviously, recapitulation is used to combat the natural extinction of learned material by progressive forgetting. School-days, whether or not the happiest in life, can make a fair claim to being the most varied and demanding in learning stimuli that are ever met. Since the previous science class these children will have been mentally shaken up by a whole kaleidoscope of lessons with a host of brand-new items crowding out the facts that we had so carefully implanted in their minds. This must be one major excuse for the discouraging hesitation which can greet the mild opening question, 'You remember what we were doing last time?' Not by threat of punishment nor by inducement alone, but by every craft in the teaching profession that we can muster, the beleaguered schoolchild needs to be helped to remember.

Forgetting is a natural process. Research has shown that it can be related to a lack of organisation, to the lapse of time, to the speed of learning, to the number of extraneous stimuli and to the infrequency of reinforcement. None of these factors seems very surprising. Among many other theories there is a mathematical model of memory which contrives to follow the general pattern of forgetting, as displayed by recall tests, with some success.[6] This assumes a two-tier mechanism within the brain, a limited short-term memory from which there is a comparatively swift leak, and a long-term memory bank to which the information can be passed after some 'processing procedure', where it may remain almost indefinitely. Although the existence of only two such memory states is not so clearly apparent, something of this general kind can be observed. After the lesson is over, information-decay, or forgetting, sets in and seems at first sight to follow the familiar decay curve so often to be met within science itself. The final conditions of maximum entropy correspond to completely disordered or lost information. However, suggestive questions can usually trigger considerable recall and a short retraining session will change the forgetting profile dramatically. Not only is the rate of subsequent forgetting much slower, but the final score of remembered items does not seem to tend to zero with a lapse of time in the way that a pure decay theory might predict. Perhaps we have aided the elusive processing procedure. There is evidence of the existence of this long-term memory bank from neuro-surgery, where vivid memories of youth can be evoked by electrode stimulation. About a less drastic method of retrieving items from this deep store theory is disappointingly silent, but the common experience of teaching insists that frequency of use vastly increases the continued availability of learned material.

Figure 6.1

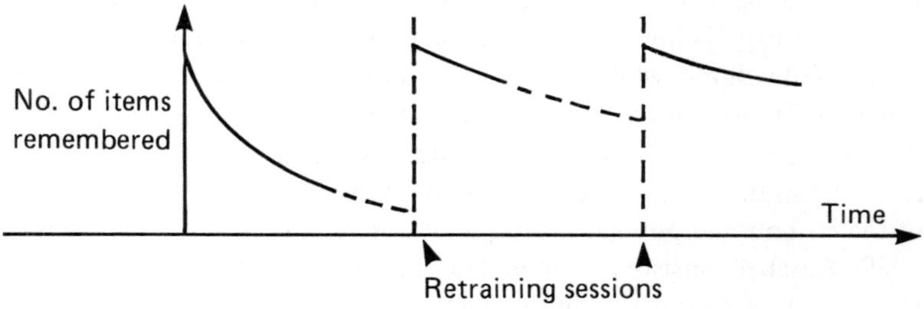


**Recapitulation through Tests**

The most obvious way of reinforcing information is by a quick resumé of the salient points from the last lesson at the beginning of each new one. The particular method employed will depend on the behaviour of the children and the measure of activity which we wish to generate. A very quiet class may listen docilely enough to the teacher reciting the important points, but this is both soporific and provides no feedback on how much of this information has been absorbed. A brisk round of questions will raise more activity in an attentive class where young pupils will often raise their hands with straining eagerness to be the first to answer. A restless overactive class must be quietened at the very beginning of a lesson and a short piece of set writing – even some straight copying from the blackboard is more effective for this purpose than any amount of angry calling for silence. Even written tests can be fun if the drill is fast and the score is high – everyone enjoys success! When necessary the scope of the questioning can be broadened to include items proved to be poorly retained in the previous test. This method is an invaluable guide to our own teaching efficiency as well as providing some insight for the children themselves into their powers of retention. My personal memory includes items of Latin vocabulary which featured in weekly tests some thirty years ago, thus ensuring that phrases like 'quam celerime' remain indelibly printed in the mind when useful telephone numbers of current friends obstinately escape me!

**Back to the Foundations**

This recapitulation has a positive, forward-looking purpose as well as the more obvious negative one of preventing the extinction of learning. The material absorbed during one lesson may be vital for the understanding of the next. This is reflected in the art of curriculum design. Density follows on the understanding and measurement of weight and volume, while reproduction and sex education are based on some familiarity with simple plant and animal cells. This is well demonstrated in the organisation of the new science courses, as well as featuring in the educational philosophy of David Ausabel.[7] In the teeth of the current and popular mystique of pure child discovery during the 1960s Ausabel consistently defended 'meaningful expository learning'. He elaborated on this idea by referring to existing 'subsumed' concepts in the student's mind and to 'advance organisers' built into the learning schedule which act as cognitive bridges for the efficient assimilation of new material. It is relatively easy to fashion such a network of cognitive bridges and many of the modern science courses publish elaborate flow-charts to illustrate the continuity of their sequences. Sometimes the very complexity of these charts tempts one to suspect that a detached aesthetic enjoyment of the exercise in itself has seduced the organisers into a private merry-go-round of alternative rotating loops.

For the purposes of continual reinforcement the straightest, most logical route from concept to concept is not always the best. A path that doubles back to allow for the reiteration of earlier, half-forgotten material or the comparison with a similar topic met in another context weeks or months before can be very valuable for refreshing the memory. Consider, for example, a sequence on gases in the air followed by combustion, respiration and photosynthesis; logic would suggest that the rusting of iron should take its place next to combustion, following the burning of candles, magnesium and copper. A mere act of oversight which necessitated squeezing in this topic at the end of the sequence showed what unexpected value this had for refreshing the earlier learning. Respiration and photosynthesis had used up a considerable span of teaching time studying the structure of the lungs, the functions of leaves and the utilisation of energy in food chains. The initial work on the composition of air had already been weakened by forgetting and a study of rusting now provided a splendid opportunity for revision. This curious circumambulation through topics is reminiscent of the 'spiral syllabuses' of Bruner and other early curriculum innovators, but its purpose here is not to give progressively wider overviews of science, only to aid efficient remembering by recapitulation.

This leads naturally to another possibility for recapitulation based on wide curricula. The newer syllabuses in science are all interdisciplinary – the Scottish Integrated Science, Nuffield Combined Science, SCISP (Schools Council Integrated Science Project), LAMP (for the Less Academically Motivated Pupil) and Working with Science. The detractors of this modern trend claim that it would take a latter-day Socrates to be well enough versed in all the sciences to teach such courses really effectively up to O level. Certainly biologists and physicists tend to view each other's provinces with some trepidation, and even the first three years of a combined course often entail a measure of tutoring among the teachers themselves behind closed doors after school hours. On the credit side it is a great help to the harassed makers of the school time-table and, more importantly, it enables teacher and pupil to gain a much greater familiarity through more frequent lessons. This blurring of the traditional boundaries also provides a new dimension for reviewing a concept from different points of view within the course. This is not to take up a dogmatic stand on the vexed question of whether science can be taught as a single discipline. Indeed it is essential that the different approaches of the traditional sciences appear within the teaching scheme and the pupils often pick out one or other of these regions to match their own preferential interests and ways of thinking, whether or not they have been labelled by name. Nevertheless the actual practice of teaching 'integrated science' provides stimulating possibilities for cross-reference from different angles. It does violence to the nature of scientific knowledge to pretend that the Principle of Moments is to be taught in the same way as the structure of the human ear, and yet the lever action of the ossicles of the middle ear provides another opportunity for reinforcement. Every time a topic is mentioned again, and its usefulness is stressed, the memory bank is forced to reshuffle its contents and loosen its grip on some buried material. What appears at first sight to be an unproved ideological point about the holistic nature of science turns out to contain, at the very least, craft-teaching value.

Demonstration, inquiry and recapitulation are only three possible methods of beginning a lesson – no doubt there are many more. Before the main work of the lesson begins, this scene-setting stage is of vital importance. It is here that we attempt to shape the restless crowd of children who enter the laboratory into a viable class of attentive learners. We may manage an individual greeting to some of the early arrivers and we will learn to recognise and value all the different personalities before us, but it is the whole class that must learn. For a little while at least we will need to hold their eyes and their minds in total attention before the next stage of the lesson begins.

## Notes

1. A. Storr, *Human Aggression* (Penguin, Harmondsworth, 1971).
2. D.W. Winnicott, *Collected Papers* (Tavistock, London, 1975).
3. W. Romey, *Inquiry Techniques for Teaching Science* (Prentice-Hall, Englewood Cliffs, New Jersey, 1968).
4. *Nuffield Physics Teacher's Guides.*
5. The Nuffield Combined Science Course for ages 11-13 is rich in examples of this sort. Their 'Activities' often provide starter experiments of exactly the kind that are needed – 'Feeling sounds' or 'Tasting acids' (acid drops). Most are very effective, and only a few suffer from over-writing in place of the more purposeful instructions which would give the activity the style and dignity of a 'real' experiment.
6. See, for example, R.C. Atkinson and R.M. Shiffon, 'Human Memory' in *The Psychology of Learning and Motivation* (Academic Press, London, 1965).
7. See, for example, David Ausabel, 'Cognitive Structure and the Facilitation of Meaningful Verbal Learning' in *Readings in the Psychology of Cognition* (Holt, Rinehart and Winston, New York, 1965).

# 7 DISCUSSION AND PREDICTION

## Theory-laden Experiments

The natural progression of any science course is signposted with important experiments. This will include familiar, significant items such as the neutralisation of an acid with an alkali, enzyme action and the relation between force and acceleration, to name but a few. They are key experiments which use, but should not introduce, basic scientific concepts and therefore cannot figure at the start to any topic in the syllabus. The most obvious point about these experiments is that they cast light upon some important mechanism, and so underpin a scientific model or theory which will be widely used in further work. We will need these experiments to give positive, unambiguous results and to carry real conviction; to this end their presentation and interpretation will be at least as vital as their practical details.

The teaching of these seminal experiments is dogged by two ideological arguments which actually muffle their impact. The first of these is the hoary educational belief, reinforced perhaps by the writings of Schwab, that experiments should be understood, if not actually rehearsed, in their historical setting. Looking back, we can appreciate their enormous rhetorical power and so, it is argued, our pupils will be enabled to understand the scientific method as they learn the facts of science, treading in the footsteps of Joule or Lavoisier. It needs only a modest acquaintance with the history of science to perceive just how uneconomical such a course would prove. Not only do we lack the time to retrace history in this way, but it would also be clearly absurd to have to teach the substance and evidence for the Caloric or Phlogiston theory in order to show how they were refuted.

The other familiar argument is that we should use these experiments to demonstrate that the proper way to mount a new theory is by direct naïve questioning of nature. The children ask and the oracle will answer. It is supposed that the problem itself will arise out of the pupil's preliminary work, where it may be posed by the child in private argument or in a class discussion. Failing that, it is presented by the teacher in a ruminating, wondering way which is thought to resemble a child's approach. (In practice the old-fashioned direct and challenging question from one-who-knows to those-who-don't offends the susceptibilities of the adult designers of teaching courses far more than it does those of the quiz-loving children. So, fortunately, this contrived attitude of innocence is not at all essential.) This style of presentation insists that scientific answers are not to be sought either from books or from knowledgeable adults but directly by the individual exercise of inquiring experiment. The teacher then encourages this 'let's find out' approach with no more than the provision of suitable apparatus, thus allowing his pupils to act without the temptation to manoeuvre their results towards a predestined conclusion. Everything suggests that a scientific experiment, if performed with care and honesty, will provide a clear and unequivocal answer to a scientific conundrum. This, perhaps, is the kernel of the guided discovery method.

Obviously there are questions that can be answered in this way, for example, 'How much does this iron bar expand when heated?' or 'What gas is evolved when lead carbonate is treated with acid?' These questions are relatively trivial, they do not elucidate any important scientific mechanism nor introduce a new scientific theory. Given a more basic query, the verdict of an experiment is neither so easy to acquire nor so simple to interpret. Few theories have arisen fully armed from a single experiment; rather it has always been the theory, evolved internally by an imaginative act of analogy, which has suggested not only the method of the experiment but also its probable outcome. In the research laboratory and in the classroom significant practical work is more like the hunt for a missing piece of a jig-saw puzzle whose silhouette is recognisable in advance than like the blind search for what may be revealed around the next corner. In this mode prediction and preconception are entirely respectable and can be used to add both spice and intelligibility to the experiment.

## Reflections from the History of Optics

This characteristic of experiment can be well illustrated from the uncertain meanderings of historical science. Inspired by the beauty of the rainbow, the scientists of the Middle Ages spent more than three hundred years trying to solve the simple optical problem, 'What is the connection between the angle of incidence and the angle of refraction when a ray of light passes from air into water?' No question could seem more susceptible to the experimental method. The practical set-up was easy enough to contrive. A curious point of this remote piece of scientific history is the difficulty that was experienced even in the drawing up of accurate tables of angles. Again and again the premature flutterings of a hopeful numerical pattern distorted the readings that were recorded. Perhaps we should judge these fudgings of results as merely on a par with the more flagrant modern hoaxes, Mendel's over-optimistic genetic statistics, Lysenko's 'vernalised' wheat, or the pathetic story of the 'patchwork mouse'. In fact these optical angles were not then so easy to measure and the margin of error did allow some latitude for accommodating the experimenter's own pet theory. As more accurate results were obtained, it became more obvious that they fitted no proposed numerical pattern, and neither was any progress made in the fundamental understanding of refraction. It is significant that both these objectives were achieved at a blow. The story of how Descartes watched cannon balls being shot into water from different elevations and altering the angle of their trajectory at the air-to-water boundary is well known. This spectacle presented him with an analogy for the mechanism of refraction, some elementary mathematics then served to predict the proportionality of the sines of the angles. Now when he set out to measure these angles he was seeking to confirm, not a numerical hunch, but a preconceived mechanism. The fact that his model is not now acceptable, his 'cannon balls' of light had to travel faster in water than in air, is irrelevant.[1]

## The School Science Problem

With young children no one would make the mistake of setting such a difficult discovery problem as this. The aim of a school experiment is much more likely to have an 'either/or' solution, or a numerical pattern of a more recognisable type. However, even in these simple cases, the same prior understanding of a suggested mechanism and hence the implications of the possible results can lend a level of intelligent participation to the exercise far beyond that produced by the innocent gropings of pure discovery. The difficulties of organising and carrying out discovery experiments have been discussed in Chapter 2, but when it is the underlying theory which is at issue this approach is altogether misplaced. We cannot rely upon the children to invent suitable theories to order and if some of the more confident do have a view as to which way the experiment will go, it is far more likely to be based on pure hunch than on a viable, imagined mechanism. Scientific method is not, sadly, so instinctive to the young.

Consider, for example, the simple and oft-repeated experiment to determine whether a black or shiny surface is the better absorber of heat radiation. As usually presented, the problem is set by the teacher. It is an unlikely question to arise spontaneously from the idle wonderment of children, and so the pupils proceed along the well-trodden experimental track. Temperature readings are taken and gradually the background chat changes from, 'Take it now!', 'Help, where's the thermometer?', 'Did you say 27½° or 28½°?' to the more hesitant 'What did the others get?', 'Should it have been the black one?', 'Have we got it right?' However unambiguous the results may be (any class of children doing any experiment can be relied upon to produce a few impossibly wrong results by holding the thermometer upside-down or entering the numbers in the wrong columns!) the general atmosphere of unsureness is not resolved by the experimental results, but only by the teacher's final summing up. The whole project after all was his idea in the first place. What basic point could there be in showing that a black surface – or pink or blue for that matter – absorbs more heat than a shiny one?

**The 'Passion' to Understand**

This experiment has been taught in almost exactly this way to so many generations of schoolchildren that it is hard to visualise any alternative presentation. However, the criterion of any lesson is clearly not the learning of an isolated fact but the progress in thinking along scientific lines. A new approach is suggested by the dramatic change in mood which develops as the teacher tries to make sense of the experimental result. If the summing up begins by linking the word 'shiny' to reflection and 'bouncing off', some keen pupil will often jump triumphantly to the end of the argument with an eager explanation of why the result had to be just the way it was. The interruption may well be incoherent, so that we have to repeat or clarify it, but there is no mistaking the change of mood that it produces. Here is no longer the lucky experimental coincidence or the meritorious carrying out of instructions, this is the delicious childish joy of having been proved right and, beyond that, the immeasurably valuable realisation that natural phenomena, unlike stamps in a collection, can be logical, predictable and comprehensible. This must be the central credo of every practising scientist.

There exists a passion for understanding just as there exists a passion

> for music. This passion is more common in children and is usually lost with age. Without this passion there would be neither mathematics nor science. This passion which has always existed in me has never lost its sparkle.[2]

The contrast between these two moods of the class, the uneasy following of blind instructions and the pleasure of successful prediction, can be so striking that one is forced to wonder whether the whole order of the exercise should not be reversed. Would it not be feasible to begin with a discussion of mirrors of the polished reflectors behind radiant heaters in order to produce a speculation from the class about the properties of shiny and black surfaces that would sustain them later while the practical work was in progress? Obviously this would totally change the dramatic impact of the experiment. No longer would there be a patient and dispassionate waiting for nature to reveal her hand, but in its place an anxious partisanship eager to achieve one result rather than another. Is this scientifically reputable?

## Induction by Experiment

The older received theory of scientific method would certainly repudiate any such approach. In the early seventeenth century when English science was just bursting forth into its glorious age of discovery Francis Bacon laid down, somewhat ponderously, his rules for scientific induction. First, he insisted, one should gather together all the experience and experimental evidence that is available with a completely open mind, then draw up copious lists of instances and counter-instances to examine their general features, and only then will the universal rule or 'law of Nature' become apparent. His own efforts to apply this method proved sadly ineffective even when judged by the standards of the times. Nevertheless there was real enthusiasm behind his search; he styled himself the 'First Explorer', envisaged a Utopian society devoted to natural philosophy in the New Atlantis and, in a more practical vein, encouraged the kind of scientific activities which led to the foundation of the Royal Society. All this, combined with his prestige as the philosopher of the brilliant new epoch in science, ensured that his inductive view of the scientific method would last unchallenged for several centuries.

In practice there have been surprisingly few scientists who, by their own admission, have travelled this naïvely experimental route towards discovery. Perhaps the prime example of the Baconian method at work appears in the writings of Charles Darwin. In the opening paragraph of *The Origin of Species* Darwin was at pains to emphasise that it was the biological and geological facts that he had recorded without speculative bias during the voyage of HMS *Beagle* which, much later, started his famous line of thought on evolution.

> On my return home, it occurred to me, in 1837, that something might perhaps be made out on this question by patiently accumulating and reflecting on all sorts of facts which could possibly have any bearing on it. After five years' work I allowed myself to speculate on the subject, and drew up some short notes.[3]

Modern historical sholarship and a reading of his fascinating journal have cast some doubt on Darwin's assertion that his collection of evidence and reflection upon it was made with a mind so innocent of all speculation. Nevertheless this is the model of scientific discovery that has had such an appeal for educational reformers. Philosophers refer to it as 'Naïve Realism', but it would be unfair to attribute the discovery method solely to a contemplation of the history of science or the traditions of philosophy. Rather it has been prompted by the more sympathetic ideal of encouraging children to enjoy rediscovering the theories of science through their own experimental activities without clues or prompts. At the same time it would demonstrate that nature, not man, is the source of these theories which are presented as the inevitable outcome of honest experimental endeavour.

We can see this investigational method at work in most of the Nuffield courses, although their latest revised versions deny, with surprising indignation, that they had ever propounded a discovery philosophy for science teaching. A later scheme to emerge from this stable is Nuffield Secondary Science, planned for pupils in the 13-16 age range, who will *not* be sitting the O level examination. The general preface by the organiser states unequivocally: 'the importance of discovery by pupils lies at the heart of much of the work contained in the Themes.' The fourth Theme in this series is entitled 'Harnessing Energy' and begins with a circus of small experiments and demonstrations which might have warmed the heart of Bacon himself. They range from growing small seedlings to model steam engines raising loads and even chemiluminescence, some twenty different items in all, many of which are imaginative and spectacular. Introductory explanation is kept to a minimum and simple written instructions are given with each set of apparatus.

> In short, what we are doing is to present the pupils, through limited experiences, in a very short time, some of the evidence which has, over the centuries, led scientists to the formation of the concepts of energy conversion and energy conservation.

After the activity is over these pupils are supposed to find themselves as irresistibly drawn to the proposition that energy is neither created nor destroyed as were Joule and Helmholtz. With children of such average ability this would indeed be a splendid achievement!

In practice this great variety of examples tends to obscure any clear-cut conclusion. The value of this collection of experiences lies in its initial exhibitory nature (see Chapter 6) and serves admirably to consolidate an already half-formed concept of energy. As evidence for the Principle of the Conservation of Energy it has little to offer. This exact law requires the kind of quantitative examination that Joule himself brought to bear on the problem when, as a young man, the principle first caught his imagination. In this early work he measured the amount of heat generated when a known mass of zinc dissolved in acid. He then repeated this experiment with a zinc/acid cell that was driving a motor and found that the loss in heat evolved within the cell was almost exactly matched by that liberated by the mechanical action of the motor. This was no blind, unprejudiced inquiry, but a beautifully conceived experiment which would never have occurred to someone whose imagination did not already cherish a new and exciting theory that needed to be established before a sceptical scientific community.

## The Partisan Experimenter

It is far easier to find examples of particular scientists basing their new theories or discoveries on such frankly partial experiments than of those who used Bacon's unbiased inductive method. Louis Pasteur, for example, was quite explicit on this point. Every experiment or observation that he planned was tailor-made to test some guess or theory which he had already formulated: 'Without theory, practice is only routine governed by the force of habit. Only theory can breed and develop the spirit of invention.'[4] The purpose of this historical digression was merely to establish scientific credentials for a partisan attitude towards experiment. For its school use we rely more confidently upon its direct psychological appeal to children and upon the added dimension of understanding that it lends to the action of an experiment. Although it may seem to be the very reverse of discovery learning, the excitement that it raises puts it in a class apart from the traditional 'To verify that. . .' genre of experiment.

**Establishing the Atoms**

In the face of the biggest and most splendid theories of science we find ourselves in the deepest difficulty. There is not, for example, a single school experiment which can be relied upon to convince a sceptical child that all matter is composed of the 'hard, massy atoms' of Newton and Dalton, but there are many that can be relished, both in advance and during performance, in the light of its explanatory power. The difficult oil-film experiment would, logically, have to be repeated many times with widely differing spreads of oil all yielding almost exactly the same value for the thickness of the film in order to produce any real evidence. Even then there would be no necessity for the concept of a minute, indivisible atom to spring forth like Venus from the waves! It is far better to expound the whole theory verbally and with conviction – after all, it is no novel idea to any child of secondary school age – and let them argue out its use and power through a number of experimental exercises. Those which have been found most useful include:

(a) making arrangements of polystyrene balls packed into pyramids or tetrahedra to compare with actual crystals; then
(b) demonstrating the cleavage of crystals along planes that can be seen on their models;
(c) handing around Laue X-ray photographs of crystals as more evidence of internal order;
(d) having our pupils mix together carefully measured volumes of alcohol and water and then repeating the experiment with sand and dried peas.

All of this is unabashed expository teaching, but it makes a clear visual impact just because of the immediate comparison with such simple representations of the important but invisible atoms. Now we can progress to the idea of these atoms in motion through diffusion experiments in gases and liquids. This is more difficult because we can present no simple visual analogy, so we will need to sharpen the imaginations of our pupils for them to see these effects in the light of an atomic model.

(e) The diffusion of bromine vapour (discussed below);
(f) Brownian motion (see Chapter 5).

The Nuffield chemistry organisers to whom all teachers of science are so indebted for their inventive experimental approach actually pioneered many of these experiments, but still attempted to stand by the older inductive method of learning.

> The first step is the accumulation by observation and experiment of as much information as possible about as many kinds of materials as possible. The second step is the classification of these observations into convenient categories. The third step is the search for the underlying explanation. . .A school science course will proceed on similar lines.[5]

Teachers who abided honestly by this method were continuously interrogating their pupils: 'What does this experiment show us about the nature of matter, is it continuous or grainy?' As often as not their pupils sat unhappily bewildered and indeed, with such evidence, who could blame them? Happy children, like their heroes among the detectives of fiction, work best when they have a clear hunch and are busy showing that it is *right*! We need not worry that this approach will make science seem too easy or too sure, plenty of their favourite type of question will soon show just how incomplete is the total field of knowledge: 'Is there life on other planets?' and 'How do our brains think?'

## The Sequence of Learning

The hunch or partisan approach to experiments is only a part of the teaching process, one link in a chain which may involve a variety of different methods, each orchestrated to a gradually increasing tempo of learning. There may have to be an opening stage which is relatively slow in its achievement of testable results – experiences and observations along with discussion to build up a useful concept (Chapter 5). The next stage usually takes the form of expository teaching and some written notes to consolidate the learning before we can launch a discussion which will generate a pattern to look for and the prediction of an experimental result. Only then will the class be ready to learn from an important experiment. It is the consolidation and discussion phases of this scheme that highlight the contrast between it and the discovery or induction method. Much of the theory has already been taught, so the purpose of this discussion is not to generate such theory from the pupils' 'discoveries', but to enable the teacher to assess if the basis of theory has been correctly understood by talking over class ideas for testing or expanding it. This is also a good time for some practical inventiveness on the part of the children. The final experiment is now far beyond the realm of open-minded induction, it is more of an arbitration between rival ideas to whose outcome the pupils will already be fully committed.

There are some features in this instruction sequence which are similar to those proposed by the American psychologist Robert Gagné. In *The Conditions of Learning*[6] we find the 'attributes of objects' leading to concept-forming and followed by rule-learning. The final stage in Gagné's model, however, the problem-solving, serves a different purpose from the discussion of experiment in the above scheme. Problem-solving is a process by which learnt rules are applied to different situations to yield new knowledge. It leads away from the field of inquiry, whereas science requires a great measure of 'introspective' experiment to ensure that the bare bones of theory be clothed with essential detail. If heat radiation is like invisible light rays, can it be reflected, and if not, what happens to the heat? If plants can synthesise sugar in sunlight, what part of the leaf and what colour of light are responsible for this? Gagné's scheme is admirably suited to the teaching of mathematics, but when he applied the same processes to the teaching of science he failed to see the special character that experimental problem-solving would assume:

> similarly, a goal of science instruction is not simply having the student state a concrete example of a scientific principle but also making it possible for him to apply this principle in novel situations, and perhaps even to design an experiment to test a principle of his own devising.[7]

The faint strains of the discovery theme are unmistakable. One topic is being left behind and the next is being brought into focus by a combination of the stimulus of a problem set by the teacher and the unaided invention of the children. Gagné himself refers to this as the 'discovery or problem solving' stage, although he is careful to insist that there is a prerequisite knowledge of basic rules without which we can have nothing but the rawest of hopeful discovery teaching with its sad lack of efficiency.

Gagné's theory has been influential and has inspired a whole new syllabus, the Schools Council Integrated Science Project (SCISP), which openly boasts that it is the first curriculum development to be based on an explicit model of the learning process. They restate it:

INFORMATION → CONCEPT → PATTERN →
PROBLEM SOLVING

with the problem stage often ranging very far from the original pattern-seeking activity. This is particularly apt in a course which proposes to teach intellectual skills rather than facts or theories and to apply these to observing the impact of science on society – ambitious aims indeed! They make clear the framework of this model to the pupils themselves, for this is not so much a project concerned with how best to teach science as how to train the children to tackle topics beyond those actually studied. Questions posed include the siting of open-cast mining in regions of scenic beauty and high unemployment, or even deciding what factors should influence a doctor's selection of patients for an artificial kidney machine. No one would quarrel with an attitude of social concern, although some might doubt the confident reliance on the emotional maturity of the children and on so simplistic a model for universal problem-solving. Their defence is frankly combative: 'In the absence of an explicit model such as SCISP, tacit models will be used: for instance political models, prejudice models or eighteenth century models of what science is.'[8]

## 'We Discussed it Till I Knew I was Right!'

In the discussion phase of a lesson the tools of problem-solving are used nearer home, when they are turned back on the theory that is being taught to test it and examine its working in a way that is peculiarly scientific. It is a medium through which the children adapt their imaginations to the new mechanism, mentally shaking and flexing it to examine its use as an instrument of thought. Our pupils will need time to argue and invent audibly so that others can comment and make counter-suggestions. When necessary, the teacher himself can clarify misconceptions about the model which has just been taught, and this is an invaluable aid to assessing our own teaching and the intellectual development of the children. Contributions often prove fascinating and unexpected, and most teachers become involved in the discussion, encouraging the diffident, breaking up pauses with helpful (or outrageous) suggestions and cooling down tempers when the argument becomes too hot. An experiment is decided upon, but not in the vein of 'Let's see what happens when. . .', but, because they are fired with the fervour of debate and childish confidence, 'I *bet* it will. . .' Now all that is needed is a quick check that all the class know what is at stake and are prepared to back one prediction or another. Once committed, our pupils will work much harder to support their 'side' – the familiar football-fan syndrome, but also not unlike the partisan attitude of many adult scientists.

To breathe some life into these generalities let us consider the topic of elementary magnetism. All the introductory experiments, picking up nails, suspending a magnet in the meridian, and letting it act upon another magnet, point inescapably to the existence of powerful 'poles' at either end of the magnet. It would be a mistake to ignore this elementary stage, whatever our private theoretical goal. However, once our pupils have learnt how to magnetise a piece of steel by stroking it along its full length with one of these potent poles, the time is ripe for some consideration of the internal arrangement of magnets. At first sight this appears to run counter to the earlier concept of poles. Dropping or heating a magnet lends some support to this domain theory, but more is required. This makes a good launching point for discussion. Can we tell whether there are just poles at the ends or an arrangement of magnetic domains that runs the whole length of the magnet? 'Why does the middle of the magnet have no power of attraction?' At first, few of the children will realise that the alignment of the domains is itself enough to predict this lack of power, and they may make wild suggestions about 'confused' domains, or just see it as proof that poles only exist at the ends of the magnet. It needs diagrams to show what each theory predicts, and at this point some pupils may gleefully offer to break a magnet in two to decide! A similar experiment is easy enough to carry out with lengths of wire which the children can magnetise, a compass, and a pair of pliers with which to cut it. The pause and discussion here has been an important tactic, not only to ensure that the experimental procedure is fully understood, but also in order that the class can take sides and guess what may happen. Then the experiment gets under way with a speed and determination which owes much to childish competition and the result is greeted with the bombast of simple triumph: 'I was right!'

It is impossible to foretell what contributions will be made to the discussion, but these may well include hopelessly 'wrong' approaches to the problem, full of childish, animistic ways of thinking which are just as valuable for us to hear as are the confidently causal ones. The young child's world is rich with feeling that no one would wish to destroy by sarcastic contradiction, but which must be augmented, within the laboratory, by more mechanistic imagination. The same experiment, performed without previous discussion, not only has less impact but may also be completely misunderstood in an unforeseeable magical way which can be mysterious to adults. Almost every science teacher has experienced these embarrassing failures with a class that has followed through an apparently foolproof experiment, observed the correct result, but has identified a sympathetic power in place of an inanimate force. The diagram illustrates a peculiarly vulnerable experiment. It is used[9] to show the respiration of a woodlouse. The soda-lime absorbs the exhaled carbon dioxide and the consumption of oxygen is shown by the movement of the coloured water index. Despite its apparent simplicity, a considerable proportion of 12-year-olds, if asked after the experiment, will maintain that the water moved along 'in order to give the woodlouse more air'. This must indicate that, at least to some extent, the whole watching and measuring operation has been to no avail. What they 'observed' was the force of compassion rather than the rate of breathing! It is too late to correct this point of view after the experiment is over, and the whole experience would have had a completely different complexion if the causal argument had been understood before the practical work began.

Figure 7.1

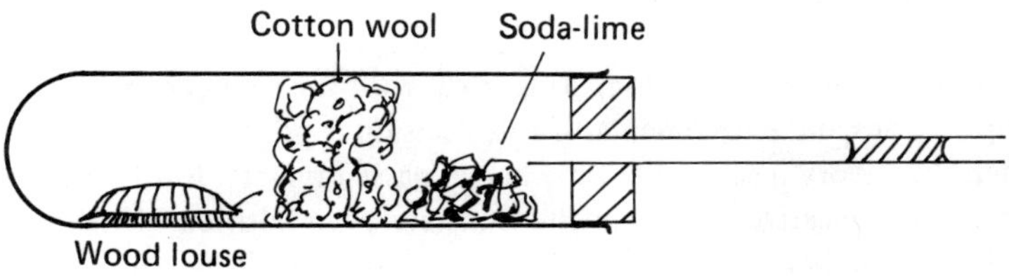

**Rival Theories and the Popperian Test**

For children to espouse an abstract idea and to make it their cause, to argue from it, become involved and yet to accept that there is a natural experimental arbitrator, and then to embark on a test fired with real motivation, albeit chauvinistic, is in itself a delightful teaching achievement. This is education for the whole child – intellectual, imaginative, sociable and disciplined. Within the sophisticated domain of the philosophy of science it claims another affinity. When a class becomes polarised by an argument between two rival theories and sets about deciding the issue by confirming or falsifying a prediction, we can see, in microcosm, an example of Karl Popper's famous Falsification Principle at work. Popper's original thinking was much influenced by Eddington's test, in 1919, of Einstein's Theory of General Relativity. Would there be a measurable displacement of stars viewed near the limb of the sun during a total eclipse, as predicted by Einstein, or would there be no displacement, no deviation of light passing through a strong gravitational field and so a refutation of the new Relativistic Gravitation? Newtonian Mechanics or General Relativity? It seemed to Popper as though the fate of these two immense edifices of thought hung in the balance. For nearly two hundred years Newton's theory had been triumphantly successful, even to predicting the existence of new planets, but, in the event, displacement of the stars was observed and Einstein's theory went forward to face other different problems, while the older rival theory came to be seen and used as a mere approximation in cases where no such large gravitational fields were involved.

It has been comparatively rare in the history of science that two such rival theories have confronted each other with opposite predictions (see Kuhn's demarcation between normal and revolutionary science) but Popper made this the cornerstone of his theory of the growth of scientific knowledge.

> [W]e seek a decision by comparing them with the results of practical applications and experiments. If this decision is positive or verified then the theory has, for the time being, passed its test. But if the decision is negative, if the conclusions have been falsified, then their falsification also falsifies the theory from which they were logically deduced.[10]

**The Power of Prediction**

There is no suggestion that our young pupils be burdened with the logic of this method. On the contrary they, and indeed most of the non-philosophical population, recognise both confirmation and falsification with equal conviction, and this too is valuable. The point is that a scientific theory should be able to make a prediction with practical consequences; it is a working model and not a mere verbal description or classification. It is not immediately obvious to young children that the question 'What is light?' is answered in a totally different way to the question 'What is a cow?'[11] This is not only because the latter is visible so that its features can be recognised, while the former, in a paradoxical but obvious sense, is invisible; but because the answer 'Light is an electromagnetic wave' is not to be understood as a mere description. It is a predictive model explaining interference, velocity, polarisation and a host of other observations. Demonstrating this working mechanism at the heart of every scientific theory is one of the chief tasks of the discussion phase of the lesson. In many biological topics, such as the transport systems in plants and animals, the point is obvious enough. The structure of vascular bundles, the alimentary canal, veins and arteries are not dwelt upon at an elementary stage just because their transport function is clearly so much more important. It can be less clear in physics and chemistry where the tiny atom can have its 'cow-like' features dwelt upon for far too long at the expense of its kinetic, working function which leads so readily to predictive experiments.

The psychological impact of an experimental test relies on the twin characteristics of (a) its results being not yet known, and (b) its drama or unlikelihood. Both these can be built into school experiments. A theory that explains familiar instances is satisfying but rather worthy; one which predicts a new effect is far more exciting, but also more convincing. It is often more worthwhile to pause in mid-experiment to elicit a forecast based on theory than to wait until the end and to use the results in retrospect to bolster a principle. The scientific community itself exhibits the same natural, if illogical, reaction to results which have been predicted in advance of the experiment. Einstein's new gravitational theory, for example, had already scored a point over Newton's by explaining the precession of the orbit of the planet Mercury which had been puzzling astronomers for half a century, but this won little attention in comparison with Eddington's public vindication of the brand new light-bending effect. Popper himself emphasised the psychological power of precise and improbable prediction. He was contrasting the tight, vulnerable explanations of science with the blanket, irrefutable explanations offered by some schools of psychoanalysis. It must be possible for the test to go wrong, otherwise we have the 'heads I win, tails you lose' effect. He wrote, 'confirmations should only count if they are the result of risky prediction.'[12] In the more memorable words of Hermann Bondi when facing up to the observational test on his own Steady State cosmology, 'a theory should stick its neck out!'

Popper's heavy emphasis on the conclusive verdict of falsification as opposed to the inconclusiveness of confirmation is not always appropriate either in school teaching or in the growth of science itself. We use discussion and the predicted experiment more to teach our pupils how to apply a theoretical model than to put the whole theory in jeopardy. As has been pointed out earlier, we use demonstrations to show *how* an atom behaves, not *if* an atom exists. Again in the field of electrostatics many children would predict that a Van der Graaf generator should light an electric bulb. If it does not, this is no falsification of the nature of electricity common to both appliances, it is a highly useful modification to the origin hunch. An unexpected result, like the lighting of the bulb only during sparking between the domes, adds clarity to the idea of current which had been vague before. The electron microscope has shown that the gaps between the fibres which make up an actual cell wall are larger than salt or sugar molecules; we do not, cannot, reject the whole life-supplying mechanism of osmosis, but modify the theory and search instead for some cytoplasmic membrane to fulfil the semi-permeable role. Even downright contradictory results have not always toppled theories – there is a stubborn, intuitive protection that science places about her most cherished hypotheses which seems almost logic-proof! Not long after J.J. Thomson had identified electricity with the minute electron Hall discovered that a few metals appeared to contain positive charge carriers. Did this cause the downfall of the electron theory? Not at all. With superb oblivion to the dictates of logic the result was simply labelled the 'anomalous' Hall Effect and conveniently shelved for some thirty years until a more sophisticated electron theory could take it in its stride.

**Imagining the Model**

School science has something to learn from this mixture of history and philosophy, if only a reassurance that spirited intellectual controversy, evolving predictive models and imaginatively used experimental evidence are the meat of real science. Much of this we can adapt to school learning. Sometimes a pause can even be created in mid-demonstration for this invaluable period of argument and prediction, as in the case of the burning of hydrogen or the diffusion of bromine.

The latter example lends itself to a particularly interesting battle of arguments without which the second part of the experiment loses much of its impact. The basic inquiry centres on how fast the bromine molecules move. Observed diffusion into air is not only very slow, but it shows a colour graduation which can be drawn and coloured in by the children to emphasise the importance of this intermingling. At first it may suggest a slow movement of the molecules gradually forcing their way through a jelly-like barrier. On reflection the pupils realise that air too must contain moving molecules. What happens when one molecule hits another? Do they bounce off backwards? Some children inevitably suggest that the molecules 'fight' each other. The animistic is gently disentangled from the mechanistic until two opposing points of view emerge – either diffusion is slow because the molecules move slowly, or diffusion is slow because fast-moving molecules keep rebounding back from collisions with other bouncy molecules.

At this point a show of hands will tell how involved the class has become by the relative number of adherents for each side. It will also demonstrate how well the young imaginations are grappling with these invisible dynamics. There is no difficulty over inventing a decisive test – the pupils will readily suggest repeating the whole experiment without the air. Now the long tube (the Nuffield apparatus is excellent for this experiment) is evacuated and sealed off before the bromine is introduced. (The simpler apparatus, using a couple of general-purpose flasks, must also be disconnected from the pump before the bromine is let in to avoid the sterile explanation that the pump is 'sucking' at the bromine molecules to make them move faster.) At the vital moment every eye should be watching avidly, partly at least because they have committed themselves in advance to the speed at which they think the gaseous molecules will move. Then when the reddish vapour sweeps up the tube or into the second flask in a dramatic instant of time the result is unforgettable. Gas molecules move fast and are 'bouncy' (*pace* the logical caution of Popper) and further experiments on the kinetic theory of gases can begin.

If the aim and order of the experiment have been so reversed it is only fair that we should be ready to accept accounts of the experiment in the same vein: 'Jill and I guessed that the bromine molecules would move like mad without the air there to stop them and when Miss opened the tap they did. It was great and I was right!'

Not only does this represent a break with tradition in the matter of writing out experiments, it is also at variance with the accepted scientific format. In view of this presumption it is a comfort to quote from a witty BBC lecture by Dr P.B. Medawar, 'Is the Scientific Paper a Fraud?':

> Then comes the section called 'results'. The section called results consists of a stream of factual information in which it is considered extremely bad form to discuss the significance of the results you are getting. You have to pretend that your mind is, so to speak, a virgin receptacle, an empty vessel, for information which floods into it from the external world for no reason which you yourself have revealed. You reserve all appraisal of the scientific evidence until the 'discussion' section, and in the discussion you adopt the ludicrous pretence of asking yourself if the information you have collected actually means anything.[13]

## Notes

1. The results of Wildebord Snell, published posthumously but apparently predating those of Descartes, give no indication of how he arrived at his famous formulation of this law.
2. Quoted in Hilary Cuny, *Albert Einstein* (Souvenir Press, London, 1963).
3. Charles Darwin, *The Origin of Species* (J.M. Dent, London, 1928).
4. Robert Reid, *Microbes and Men* (BBC, London, 1974).
5. *Nuffield Chemistry Handbook for Teachers*, Phase I.
6. R. Gagné, *The Conditions of Learning* (Holt, Rinehart and Winston, New York, 1970).
7. Ibid.
8. Mike Lyeth, 'SCISP in Brief', Printed Centre for Science Education (1976).
9. Scottish Integrated Science Course, Unit 8.
10. K. Popper, *The Logic of Scientific Discovery* (Hutchinson, London, 1968).
11. See also infuriating winning gambit in Child's Question Game:
    First move: 'Daddy, what is a cow?'
    Second move: 'Daddy, why is it a cow?'
    Checkmate in two! Moral (if any): wait for more mechanistic opening play.
12. K. Popper, *Conjectures and Refutations: The Growth of Scientific Knowledge* (Routledge and Kegan Paul, London, 1963).
13. D. Edge (ed.), *Experiment* (BBC, London, 1964).

# 8 APPARATUS AND PRACTICAL SKILLS

There are many varieties of practical work that can be used in the teaching of science and some of them come readily enough to young children. Indeed when we look at the exuberant experimentalism of Davy or Franklin or at a middle-aged professor of entomology who pounces with triumph on some new form of insect life we tend to use the indulgent epithet 'childlike'. Are we really so deceived? There is obviously a range and depth of skill involved in experimental work which demands a serious apprenticeship and deserves, perhaps, more attention and respect than it often receives from science educationalists. Here then is an opportunity to examine practical work in a school laboratory, not so much as an aid to teaching theory but as a discipline in its own right.

## The Trend away from Practical Examinations

Hand in hand with the successful promotion of modern investigation-loaded syllabuses has gone a steady *decline* in the number of practical public examination tests. There is a reason for this paradox. Current thinking about school science teaching has not only increased the quantity of practical work, it has also changed its character and aim. Where before there were techniques to learn, dissection, identification of specimens, the analysis of a mixture of salts or the measurement of resistance, these could easily be tested in a direct way during a two- or three-hour practical examination. If, on the other hand, experiment is conceived as a creative discovery activity whereby our pupils seek to uncover new rules, confirm an intuitive leap or illustrate a concept, it becomes quite unsuited to examination testing. The situation is similar to that in English or music examinations: the pupils' powers of analysis and comprehension can be fairly assessed, but to demand the creative insight for an original poem or sonata to order on a specified day is clearly absurd. When experiment is seen in this light the decline in examinations becomes inevitable and may also be a measure of the decline in prestige of the practical skills.

The various modern courses take up slightly different attitudes towards this issue, but the most forthright criticism of practical examinations comes from one of the latest of the Nuffield A level courses:

> Of one thing the physical science group were quite sure – *a formal practical test is not a desirable ingredient of an examination at this (or perhaps at any other) level.* This decision meant that whatever practical work was done, could be completely integrated with the overall development of the course; there would never be any question of doing experiments as a drill for a practical examination.
>
> What other purposes are served by practical work? Its main function can hardly be to teach students practical techniques, since far too little time can be spared in a sixth form course for any one type of experiment or measurement, for students to become really proficient in it.
>
> Its *principal* purpose must surely be to give students something of the flavour of scientific discovery. . .On the whole, it is probably better for students to carry out a large number of fairly rough experiments than a much smaller number of 'precise' ones.[1]

This seems to be a fair summary of the general Nuffield attitude towards practical work. The emphasis is on discovery and estimates of order of magnitude rather than on the cultivation of experimental skills or respect for accuracy.

The special aims and objectives of SCISP lead to a similar attitude towards practical work. They insist that the task of experiment is to search for 'patterns', a word carefully chosen to include such overall statistical trends as those between the weight of fertiliser used per acre of farm land and the gross national product, as well as a mathematical relation like Boyle's Law. This integrated approach maintains that both be accorded the same scientific status as patterns revealed by inquiry into measurable factors. Apart from the very real differences between a trend observed in unique world-wide data and measurements taken during laboratory tests which are infinitely repeatable, it is clear that this viewpoint can very easily breed a deplorably uncritical attitude towards experimental data and its statistical 'scatter'. Accustomed to the statistical results typical of sociology and economics, there is little incentive to strive after the kind of accuracy required for the corroboration of a numerical prediction of physical theory. It is not surprising, therefore, that although their final examination grades include a teacher assessment of the 'interest' and 'scepticism' displayed by pupils during their practical work, one looks in vain among their list of objectives for any mention of training for experimental skill.

**Experiment – Real or Instant?**

There is a level at which the arguments of the SCISP and Nuffield promoters make sound educational sense. The happiest mode of teaching always draws upon the natural, appealing activities of children and helps them to learn through these. Simple experiments which lend an air of discovery to topics which have to be learnt are fine and, for the consolidation of concepts, they are invaluable. With the aid of special, purpose-made apparatus such experiments may yield rapid approximate results with little technical difficulty, but, like the 'convenience' foods of the modern supermarket, a continuous diet of such commodities deprives the recipient of a full range of quality and flavour. Scientific experiment is a pursuit in its own right. School practical work is not only a mode of teaching theory, it is also our pupils' introduction to the reality of experiment with all its qualities of painstaking care, inventiveness and precautions. Science accords the highest respect to the devising of a new measuring technique, a particle-counter or a method of dating trees, and also to the skill in their use which serves to push back the limits of accuracy in our exact numerical knowledge. This work is every bit as much the substance of 'real' science as is the cerebration that produces a resounding new theory, and our pupils should be aware of this.

The appeal of this sort of work is not beyond the appreciation of children who can derive both pride and pleasure from the pursuit of accuracy. On the contrary, dissatisfied pupils will often attribute the disenchantment that they feel with a science course quite precisely to the superficiality and vagueness of their experimental results.[2] Robert R. Sears of Stanford University[3] once divided the pleasure principle in learning into 'process pleasure', as in the manipulation of crazy Emmet-type machinery, and 'product pleasure', as in the solution of a problem or the construction of a worthwhile model. Within the field of school practical work the same division exists; there is fun to be had from a quick investigation of the modern 'circus' type – watching the reaction between calcium and water or guessing the contents of a sealed container, but there is little possibility in this for skill or expertise. On the other hand, there is also a deep satisfaction to be gained from a series of careful manipulations yielding a smooth graphical curve or a visibly exact point of neutralisation.

Finally, it is doubtful if we educate well without giving some inkling of the sort of accurate work demanded in many lower- or middle-grade scientific and technical jobs. This is not a suggestion that we offer specific technical training, only that we monitor the course of experiments to ensure that they form a growing experience of practical challenge for our pupils.

## Teaching Practical Skills

The most interesting of the introductory courses in this context is the Scottish Integrated Course. Here we see practical work designed not only for the enjoyment of children, under such charming titles as 'Bonanza' or 'Experimental Circus', but also for the cultivation of such essential skills as 'Observation', 'Drawing Conclusions' and 'Controlling Variables'. Whether or not they always achieve these objectives, here at least is a thoughtful beginning and a spring-board for the examination of experimental training.

When the youngest pupils first enter the laboratory and gaze expectantly across the unfamiliar benches they receive, in this course, the very set of apparatus and instructions that they might have dreamed up for themselves in their rosiest moments of anticipation. Here are the racks of shiny test-tubes and the bottles of exotic labelled liquids. Their only task is to select two bottles from the shelf, to pour a little of each into the same test-tube, and then to watch the result! If the reagents have been well selected, many of the effects may be satisfyingly sensational in the comic-strip tradition of foaming liquids and brilliant colours. There is a table drawn up for them on the worksheet (which receives rather less attention) on which to enter their observations. Altogether this proves a very popular beginning to school science, and indeed one which has subsequently proved rather difficult to live up to.

The serious objective of all this fun is simple and direct observation, the most basic of the technical skills that we need to foster. Two comments arise from watching such activities in progress. The first of these is not about techniques, but about the lack of scientific purpose. The names of the chemicals are heart-breaking to spell for such young children, difficult to pronounce and not valuable to memorise, and it will appear at the end of the day that in spite of this effort no conclusion has emerged. This is not to devalue this delightful opening experiment, but just a suggestion that the chemicals be so grouped that some simple pattern will emerge. For example, sodium carbonate gives the same blue precipitate with all solutions of copper salts, but white with calcium salts. An almost uncontrollable urge to mix up all or any chemicals that they can lay their eager little hands upon is an aspect of original sin that merits the strongest discouragement in any well-run laboratory and we must not seem to encourage it. Rather we need to emphasise the purposeful pattern-seeking nature of science during each and every experiment.

As for the observation itself, if we have it in mind that the purpose is not just momentary attention but to train the powers of observation, there is need for a graduation in complexity once the mixing routine has been mastered. Delayed reactions such as the formation of iodine-starch blue in the Harcourt-Essen reaction, or the slow appearance of colloidal sulphur when sodium thiosulphate is acidified, are just two of the possible extensions. All skills, whether those of the craftsman or the child, are the outcome of graded exercises requiring mounting attention. Even the possibility of missing the vital moment of transition adds excitement and an element of competition to the experiment.

The Nuffield courses do not ignore the importance of experiment, far from it, but their approach is quite different. In the place of experimental training situations they sponsor the provision of initial exhibitions (see Chapter 6) where the very variety is expected to trigger off incentive for close observation. In some cases it does so. There is a restrained quality to this kind of teaching which relies on spontaneous catalysis beyond the contriving of a teacher, the serendipity of object and beholder, which is on a par with their faith in the discovery method for learning theory. Practical skills, however, are not learnt by sudden illumination but by patient and careful use on the job.

At the outset observation will be the direct result of the unaided human senses, but very soon these will need to be extended by the mediation of simple measuring instruments, the ruler, thermometer, weighing machine and stop-watch. Here is a new opportunity for training, and indeed a vital one. The accuracy of measurements made with these tools, however insensitive in themselves, continues for many years to be a function of the pupil's growing skill. This suggests that the programme of experiments we design for these early years shall involve their frequent use in increasingly demanding situations. It is also an opportunity to clarify our own thinking on the concept of measurement within science and the part played by instruments and apparatus in this.

## Do Instruments Complement or Replace the Senses?

It is possible to see an instrument as a simple extension of the human senses. It is concerned with the same perception of qualities but is more convenient and apt, as the metre rule is more consistent than the paced-out yard and the thermometer is wider-ranging and less painful than the probing finger! With this approach the scale on the instrument becomes no more than subdivisions of the grosser human sensitivity – supra-human in its powers of discernment, but not inhuman as a computer might seem to be. There are teaching advantages to this point of view. It allows the design and choice of a measuring instrument to fall into place as simply as the picking up of a sheet of fine print for closer scrutiny or the cupping of a hand to the ear to pick up faint sounds. The *meaning* of the stimulus has not changed with the invention of more sophisticated means for recording them.[4] Only with the measurement of quantities of which the human senses are not directly aware – electric current or pH value, for example – will the instruments begin to impose their own interpretation and quantification. Perhaps there will also be a moment of logical misgiving when such instruments are used in turn for the appraisal of the human senses themselves, such as defining the range of audible frequencies, but by then a chain of interlocking concepts stretching from the straightforward hearing of notes to the frequency of vibration will have made the extrapolation to 'unheard' sound a metaphor well within the sensory vocabulary.

If the teacher is already imbued with an image of science whose austere truth stands above and beyond the limitations of the human senses of the observer, such an approach to quantities and instruments may seem simplistic and indeed wrong-minded.

The conventional viewpoint on measurement is more remote. It holds that the human senses are fallible while the science that we aim to unravel is not. Like the rhythm of the stars science lies beyond the human grasp of span and duration, so we have to rely upon instruments which, by some property of mechanical objectivity, cannot be so easily impressed or deceived as are our simple senses. This, as we shall see, is exactly the approach adopted by the Scottish Integrated Course. It begs a lot of questions about the nature of these transcendental quantities that defeat our sensory understanding and yet reign supreme in the quantitative world of science.[5] Although such philosophical queries are not likely to arise in the classroom, any divorce between 'feel' and measurement does little to encourage a confident recourse to a simple instrument when the perceived signal becomes too faint or imprecise to give unambiguous results. More importantly, perhaps, such inhuman sensitivity tends to break the vital link between the quantity being measured and its sensorimotor understanding which we have been at such pains to develop (Chapter 5).

**Measurement, Feeling and Consistency**

The polarisation between these two views may be hard to resolve amongst adult scientists, but for our young pupils the former approach to measurement has some obvious practical advantages. If tools can be regarded as natural extensions of the human senses there will, at once, be an understanding – even an affinity with them in much the same way as the mechanic may regard his favourite wrench as a complement to his own felt powers of twisting and turning. The outstretched thermometer 'feels' temperature like the elongation of a finger tip, not as a deplorable pandering to animism, but as a stage in the extension of our private sensation to the rising of a thread of mercury. Even the most serious objection to this method of introduction to measurement, its subjectivity and variation, can be turned to teaching advantage. Early in any course we shall want to refer to the fixed units of measurement showing how scales can be calibrated, compared and standardised. This invokes no transcendental objective quantification nor yet a concept entirely dependent on a stipulated measuring procedure. Instead we achieve a standardised inter-subjectivity of scales, which is not unlike an agreement on perceptions, to produce the consensus and consistency that scientific work demands.

Such a discussion of measurement translates very smoothly into elementary laboratory work. Children of this age have rarely developed beyond the 'concrete' level of thinking and must rely on their activities to understand the value of measurement; as they begin to study science they will perceive and measure interchangeably. The Scottish Integrated course is unusually specific about the need for measurement in their second topic, 'How reliable are the human senses?' There is a set of optical illusions to emphasise how fallible are our estimates of length and beakers of hot and cold water to deceive our sense of touch. The moral pointed from these and other exercises is that our simple senses are *not* to be trusted and 'instruments are necessary to overcome some of these limitations' (Teacher's Guide). This is a stern, uncompromising start, but these activities, many of which produce much fun and competition, point just as clearly to the other approach to measuring instruments. As controversy arises among the children about which line is longer or which water is hotter then, as simply as a referee is called in to judge an argument in sport, the instruments are brought into use. Checked and calibrated they now decide the issue, not because the human senses are rejected, but simply for the consistent results that they produce. Any teaching which attempts to undermine the children's trust in their own sensory perception may find itself not only confounding the basic meaning of the measured quantity but also the proper province of the instrument. Should we really use the possibility of sensory deception to pour scorn on the mother who feels the temperature of bath water with her elbow, or tests the milk in a baby's bottle on the underneath of her wrist? These are the clearest possible demonstrations of the term 'blood-heat' carrying far more significance than either '98°F' or '37°C'. The intrusion of a thermometer into such an intimate situation can add neither useful sensitivity nor scientific consensuality.

## The World of the Microscope

The scientific instrument which stands to gain most when seen as a natural extension of the human faculties is the laboratory microscope. Either it can be an astonishing and delightful elaboration of the simple sense of sight, or a difficult and mysterious instrument revealing an incomprehensible field of unrelated detail. The adjustment of the mirror, the size of the aperture, the distracting image of distant windows and the difficulty of focusing all make the early lessons with a microscope almost prohibitively difficult with a large class and a single teacher. From every point of view, the confidence of the pupils, the smooth conduct of the lesson, the enjoyment of the microscope and the sanity of the teacher, those nightmare sessions when ten or more querulous voices are clamouring for help at the same time must be avoided! It does not help that the children have drawn large labelled diagrams or have copied down a numbered list of instructions. The leap from simple seeing is too great and the skills required too many. Only if the process of peering more closely at a small object – the torn edge of paper, the print on newspaper or a dead fly – can be gradually extended through the use of a hand lens and so to the low-power objective will the examination be comprehensible. A class examining the parts of a flower lying on black paper quickly learn to use a hand lens. Now the same objects simply scattered on a slide under low-power with *top-lighting* present the same kind of field of view in greater magnification, a smooth progression, without adjustments to mirror or aperture. The colours of the flower remain recognisably the same as when seen with the naked eye, but pollen grains become visible while the stamens and stigma assume wonderfully gigantic proportions. It is true that their thickness may involve different depths of focus but this presents little difficulty at low power.

This is very different from the traditional first lesson with a microscope which often involves a stained fragment of onion skin complete with coverslip and under stage illumination. When, at last, the outlines of the unimagined cell structure do materialise, usually with the help of the teacher, surprise and relief are no good substitutes for the independence and confidence that we should hope to achieve. Only when some object, like a moss-leaf, is already focused under top light will a flick of the mirror demonstrate the benefits of transmitted light.

Slightly older pupils confronted by the need to interpret thin sections have another hurdle to cross. When are parallel lines evidence for tubes or layers? Can a pattern of different size circles be formed from a cluster of similar cells? Again it is only by referral back to simple eyesight that the children can use their latent understanding. Given a thick moist plant stem, dandelion or thistle, their own slashes across and down the stem made with a moderately sharp scalpel show very easily the ready 'bleeding' from the cut ends of the vascular bundles and the long clean outlines of the same tubes when the stem is slit longitudinally. The use of the microscope comes afterwards. Hand and eye together have first to build the essential expectation needed to translate the flat faint geometry of a section into the three dimensions of plant machinery.

## Reading the Scales

The need for instruments is only a preamble to their skilful use. Every mistake imaginable will be made while the children learn. Measuring cylinders will be tilted, thermometers will be removed from the liquid to be read, oscillations will be timed from a count of 'one' and rulers will be peered at from the side. Appreciating range and sensitivity is another skill to be learnt from use. Even sixth-formers can find these concepts difficult to master but they arise quite simply in the earliest years. The elementary weighing machine is likely to be of the Butchart lever-arm type which commonly has two ranges marked out on the scale

Figure 8.1

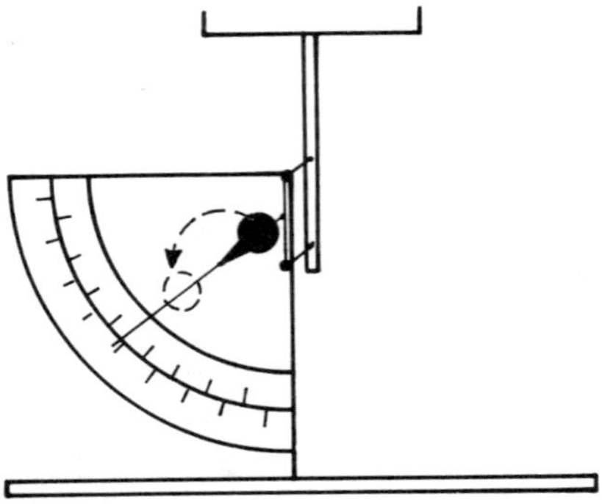

to correspond with the two positions of the counterpoise. In the third or fourth years ammeters and voltmeters will be provided with more choices of scale. These give a splendid opportunity for thought. Why use the smaller scale at all? Just deciding what each subdivision means requires reflection and some calculation, and it teaches the meaning of scales far more effectively than the old exercise of marking out an unmarked thermometer.

There is no way to understand the working of a machine more effective than making one for yourself. The Nuffield courses originated the now famous micro-balance made from a single drinking straw, but such dramatic increase of sensitivity is paid for by the delicacy of manipulation required. In average-ability classes such manual skills and the necessary patience develop later than they do in the selected groups for whom these courses were first intended. Understanding the minute fractions of a gramme involved and the arithmetical skill required for subdividing an awkward scale can make this a very difficult exercise for the first-year pupil. A home-made weighing machine with a rubber band or a floating bottle, though unimpressive in their accuracy, give their inventors a proprietary pride which is of more value than a class effort at high sensitivity with a correspondingly high failure rate and no vestige of personal invention.

**Black Boxes**

At the other extreme from the makeshift lie the glossy incomprehensible 'black boxes'; these are instruments to be used without question or curiosity merely for the readings that their pointers and dials so conveniently present. There was a time when attitudes of great vehemence were taken up on their behalf with admirers of the old Cavendish tradition of 'string and sealing wax' on one side arguing that understanding and invention were destroyed by the decadence of bought sophistication. On the other side stood the modernists trying to bridge the credibility gap between such 'Heath Robinson' laboratories and the furniture of real technology. Even British science has dropped this attitude of inexpensive 'making-do' and schools are provided with microcomputers, electric digital balances and pH meters. The original battle is now over.

There are two conflicting reactions to such technology which are as clear to see on the faces of children as in the minds of adults. Some eyes glisten at the mere sight of flashing lights and pointers, others hesitate and cringe; they have been taught to keep their hands off all expensive machinery – will they damage it? will it explode? From such fearful beginnings grows the anti-technological attitude which is so beautifully described in *Zen and the Art of Motor Cycle Maintenance*. What should be the teaching strategy?

We decide on the basis of expediency and overall comprehension. There can be few tears shed for the days when hapless pupils, equipped with bent and often useless forceps, struggled to pick up illegible triangles of aluminium in an effort to weigh to the third decimal place! Experiments on specific heat capacities or molar solutions now proceed apace with the new read-out top-pan balances and pocket calculators. There is rather more doubt when the simple siren is replaced by an electronic audio oscillator which demonstrates the increasing rate of vibration less clearly than its predecessor.

The oscilloscopes can be at a still further remove from understanding. Of course it is a favourite toy with the first or second year, who love to shout into the microphone and watch the trace react – heaven alone knows what they think is going on! The inculcation of early confidence in technology may be the aim, but it can be dearly paid for in continual adjustments required when adventurous fiddling with knobs loses the trace. At a later stage the story will be different. Once cathode rays have been taught, the effect of DC and AC on the Y plates will be predictable and a very slow sweep on the X plates can then demonstrate the purpose of the time-base as it gradually stretches out wave-forms. Effective understanding of the instrument and its intelligent use require no more detailed understanding. Nevertheless there is a further stage in their use. A sixth-form student could be expected to plan a 'systems' approach to their circuitry, for the cabinet model is now too easy and too restricted. It is a real advantage to supply our pupils with separate sets of tubes, power-packs, amplifiers and time-base modules for them to link up for themselves as the job demands. (It is surprising how many pupils seem to imagine that an amplifier can work without a source of power!) The modules themselves may be no more than 'black boxes', but the parts they play are clear. In the electronics topic of the Nuffield A level physics the basic module is similarly presented as a 'black box'. Once again the aim is a systems approach, this time to feedback loops. So long as the pupils' natural curiosity is not frustrated by such effective concealment of the vital chip or transistor, such modules make an ideal introduction to inventive applied electronics.

## Machinery in the Laboratory

There is another influence on the style of apparatus due to the latest science/technology confrontation within education. A genuine desire to add social realism to our teaching has been sharpened by recent politics. In 1977, a year that witnessed a sharp decline in the national economy, bringing with it painful cuts in spending on education, the 'Great Debate' was also launched. This vast exercise in media involvement exposed both variable standards of literacy and a lack of technical purpose in education. For some time there have been signs of parallel developments in new if rather tentative syllabuses from within the profession.[6] While we wait for these to mature it is useful to take stock of our existing apparatus from this vantage point. Is it at a hopelessly far remove from real machinery? The most obvious victims in such a first sweep would be the fragile aluminium pulleys threaded with thin and tangled string from which our pupils attempt to construct models of load-bearing block-and-tackle. To teach of 'mechanical advantage' when such feeble and unstable systems hang precariously from teetering retort stands is an absurdity! Only the slow upward haul of a truly imposing load can catch the spirit of real mechanics. Similarly, need small Meccano cog-wheels always take the place of the reality of a car gear-box, and the rat or dog-fish substitute for the more familiar carcasses from a butcher's shop? Above all we have the microprocessor.

Not as a programmed teaching aid but as a real piece of hardware it demands to be used in the scientific manner either to simulate results from theory, draw displacement/time graphs for different types of motion or to display distributions in large samples from pupils to sea shells! The microcomputer must become as familiar in the laboratory as is the calculator.

## The Practical Objectives

Once the armoury of apparatus is chosen and housed about the school laboratory we still have to consider the goals of the practical work that we intend. As Dr J.F. Kerr pointed out, 'Overt consideration of the nature and purpose of practical work in science teaching is exceptional.'[7] Most writers of syllabuses go no further than recommending its use for the 'illustration' of scientific principles or some similar ancillary role,[8] but this is not only to deny the craft its own identity but also to do a gross disservice to the cause of science itself. There are real objectives which are practical not only in their method of attainment but also in their goal. These might be listed as follows:

(1) a feel for the right instrument or apparatus;
(2) a capacity to design and criticise experiments from the viewpoint of variables, controls and reduction of errors;
(3) a chance to appreciate skill and increasing accuracy;
(4) room for individual initiative and personal delight.

## Right for the Job

The first of these objectives can only develop slowly as the vocabulary of instruments increases. It is important that new apparatus is not always handed out before the need for it has been appreciated, even in a stipulated experiment. At first it will be not so much a question of choice or invention as the simple recognition that special circumstances have arisen. The ruler must be proved to be insufficiently accurate before the vernier is introduced, the spark discharge-wire should be seen as a step towards quantification before the Geiger counter makes an appearance. The first determination of the acceleration of gravity will involve dropping a block of wood out of the laboratory window and timing its fall with a class-load of stop-watches. The children usually enjoy this hugely. Probably the slightly manic act of releasing the block and the resounding crash and possible fragmentation below are all pleasantly liberating! However, the variations in measured time and public controversy between the pupils soon make the point that unavoidable errors in this method of measurement are bringing the whole experiment into disrepute. Even before the scalar-timer is demonstrated, the class will have some idea of the type of apparatus required from having identified the weak points in the procedure. Not only are the flashing counters impressive and the increase of accuracy several orders of magnitude better, its electric 'start' and 'stop' gates make it clearly right for the job. Now the capabilities of this timer are mentally filed away as a potential tool for the future. Ideally no instrument should be used only once, the next time an experiment is being planned and discussed the pupils, not the teacher, will be in a position to call upon its special advantages. Every new instrument introduced in this manner increases the possibility for the exercise of pupil choice and appreciation of the finesse of measurement.

### Are We All under Control?

Control and isolation of variable factors within an experiment constitute so important a lesson, especially in a biological context, that even in this era of brisk discovery learning it has not been neglected. It is all so like the suspect TV commercial with its 'proof' of the amazing efficiency of one product as compared to another that our pupils find little difficulty in understanding its method. The point that can be missed is the actual dilemma behind the 'wrong' approach that the control is designed to clarify. If the children have never thought to ask such a question the control becomes contrived and artificial. 'Water makes woollen fabric shrink so why not muscle?' is a good belligerent question and its asking brings into being the right control, which will be watched with interest. In effect it is the pupils' needs as much as the teacher's recipe which should dictate the controls. Once again this calls for the class discussion of a problem and its guessed solutions before the experiment is begun. Stories from biological history make good listening for young children at this point and open their eyes to the multitude of possible sources of intruding bacterial life. The aim is that every flask should become of equal interest, the controls are no longer 'blanks' in the background but as much concerned with solving the problem as is the flask with the 'special' ingredient.

Allowing the children to select their own variables and plan their own controls can present surprising turns of childish imagination. A class asked, 'What might affect the strength of an electromagnet? Choose one factor and design an experiment to test the effect of its sole variation' may come up with some delightful new twists. Not just the amount of current flowing and the number of coils of wire, but also the orientation of the solenoid and the length of time that the iron is left inside can be among the suggestions made. Provided with no guidelines for their imagination, cooking springs readily to mind and hence their concern with the time. Now, as they design their own test, they can as successfully eliminate this factor as corroborate another. For them this is no insignificant control but the basic aim of the investigation and at least as valuable conceptually in its negative result as are the more conventional tests on the variation in amount of current when the time of magnetisation is kept constant. Growing fungi and examining the diet of worms are two other elementary projects that can yield unexpected lines of inquiry.

An additional advantage of allowing the range of variables to come about in this co-operative fashion is that the tasks can be shared out among the practical groups. After the experiments have been done there will be a 'reporting back' session where the whole class can piece together the total picture from the results presented. Beyond the image of science that this faithfully mirrors any activity which gives value to both individual effort and to the co-operative phase that knit these results into a greater whole is of immense social value to a class of competitive and individualistic children.

## No More Dud Experiments!

The third and perhaps the most vital of our objectives is to instil a respect for the accuracy and skill of the activity. Care and intelligent persistence should be seen to be rewarded by significant results. Teenage science pupils are quite old enough to have already recognised in sport, craft and almost every other field of endeavour that though excellence is hard to come by, it is not impossible to achieve. They will not easily be bamboozled into enthusiasm for a course of experiments from which no amount of careful handling and growing skill can coax accurate and satisfying results. There are experiments, such as the neutralisation of an acid, which are vital and often require more practice than one experimental session can provide. The pupils themselves may beg for a chance to 'try again' to achieve that smooth and convincing graphical curve which signals their skill and success. Their happy pride of achievement is the teacher's criterion for practical progress.

There is more doubt about the merits of a long course of practical work on kinematics with ticker-tape timers designed to 'prove' Newton's laws of motion. These instruments illustrate acceleration and velocity in a superbly visual way, but are less convincing when used to verify a known, quantitative relation. Newton's mechanics were conceived as a remote celestial abstraction and their verdict is mathematically precise just because of this abstraction. Given real laboratory conditions and a pair of wooden trolleys, a trivial experiment followed by a long calculation can do little better than show that some 10 per cent of the original momentum appears to be lost, or worse still, has been created out of thin air by an over-generous attempt to 'compensate' for friction!

This illustrates a general psychological point about any experiment which may be expected to show that one numerical answer is exactly equal to another. While pupils are pleased enough with their efforts to determine for a specific heat capacity or a molecular weight that differs from the accepted value (which they always demand to know) by 4 or 5 per cent, equality presents special problems. One look at their faces shows that no mathematical argument about the size of experimental error is likely to satisfy them that 9.4 and 8.9 are equal. Disappointing experiments are duds and need a change in presentation.

One way out of this difficulty can be used with older children and involves deliberate comparison of the variation in the 'constant' property with that of some other variable. For the examination of the conservation of momentum one of the most beautiful and enjoyable experiments is the taking of photographs by stroboscopic light of pucks which float across glass on cushions of carbon dioxide. Although a casual examination of the photographs shows clearly the overall transfer of momentum, a careful and laborious calculation cannot be relied upon to give exactly equal values for linear momentum before and after collision. This proves something of an anticlimax after the fun of the flashing lights. However, in both this and the trolley experiment referred to above, parallel calculations on the kinetic energies of the pucks show a percentage variation which is so much greater that it underlines the central point about the conservation of momentum in a worthwhile and illuminating way.

The other approach to one 'equality' problem is to redefine the objective of the practical work. The earliest experiments with a lever (see Chapter 3) use piles of unit weights and large unit distances which can be relied upon to yield equal values for the clockwise and anti-clockwise moments – so long as a thick roller is used in place of a knife-edge and a little 'nudging' of the weights is allowed! The same is not the case with other balancing experiments, so a different brief is given to the class. Young pupils will work hard to get a high value of 'mechanical advantage' out of a simple rigid lever if that is their given objective. Indeed they will compete eagerly with each other, taking measurements of distances to stake out their claims, in circumstances where the inaccuracy of the spring-balance and uncertainty in the position of the pivot and centre of gravity would make a calculation based on the Principle of Moments both disappointing and uninstructive.

There are plenty more examples of this kind of problem in numerical practical work. They range from the quite horrible 'verification' of Archimedes' Principle with an endlessly dripping overflow-can to the difficult measurements in sixth-form physical chemistry. The teacher's experimental objective is a satisfyingly accurate conclusion, his subjects are his pupils and his apparatus is their on-the-spot reactions.

In the upper forms of the school it becomes particularly important that practical work is not allowed to suffer by comparison with the elegance of the theories currently being taught. All too easily it can become a second-class activity lacking any real credibility because of continual failure to reach an accuracy of the standard that the pupils expect. Sometimes the whole problem is sidestepped by encouraging pupils to pursue more qualitative projects of their own where they will not meet the challenge of numerical error. Valuable though such projects are for the initiative that they generate, this is a tactical mistake. Skill and accuracy are worthwhile goals which can be encouraged by setting a target of 'less than 3 per cent error' and by inspecting graphical curves for smoothness and repeatability. When this leads to requests for extra time to repeat practical exercises from dissatisfied pupils it is a clear sign that high standards and personal pride have become enlisted in the endless scientific struggle to wrest yet greater accuracy from laboratory experiment.

## Doing it my Way

To realise our last objective requires a generous application of sympathetic detachment rather than a carefully constructed programme. As we train the pupils to the respectful use of viable instruments and to an exercise of the difficult art of experimental design, control and measurement, is there still room for the individual to find within the necessary discipline enough latitude to foster imagination and the private delight which arises from total involvement and creativity? It can happen. Incidents which come to mind vary from the slow painstaking enjoyment of a 12-year-old girl surrounded by the debris of flowers, tacky paper and lenses, who was contriving a montage of the parts of a flower to some internal aesthetic of her own; to the obsessive attempts of a 15-year-old boy who was bent over a burner in a darkened laboratory softening glass rod in order to coax it into a curvature that would internally reflect a beam of coloured light. Neither activity was at any great remove from the conventional experiments from which they arose, but the happy involvement was abundantly apparent and unforced because each child had found a practical activity in which he could express his own skill and individuality.

Even the eager child who comes to science club full of an 'invention' which he wants to try out must be left alone to taste failure and surmount it in his own way. It has been very hard, for example, to watch while a large brass ball was suspended between the terminals of a battery in the hope that it would swing to and fro as an 'electrostatic bell'. It was so obvious that the ball was too heavy and the voltage too low, but what value would there have been in a 'success' which was no more than the connecting up of recommended apparatus? This project needed continual effort with only the minimum of help in the supplying of different power packs and the odd grudging suggestion. The final moment of triumph after several hours of steady adaptation was so happy that the boy literally bounced up and down on the spot!

## Notes

1. J. Spice, 'The Characteristics and Philosophy of the A level course in Physical Science', *NAS Physical Science. Introduction and Guide* (Penguin, Harmondsworth, 1973).
2. J. Wilmut, 'The Effects of Project Work on Attitudes to Science by Sixth form Pupils', *Education Research,* 15, 2. Also M.B. Omerod and D. Duckworth, 'Pupils' Attitudes to Science'.
3. J. Bruner (ed.), *Learning about Learning* (US Government Printing Office, Washington, 1966).
4. For the nearest philosophy to this psychological view, see B. Ellis, *Basic Concepts of Measurement* (CUP, 1966): 'operational significance can be given to quantity names without describing any measuring operations.'
5. 'Quantities are only defined in terms of their measuring operations.' H. Dingle, 'Theory of Measurement', *Brit. J. Phil. Science* (1950).
6. See, for example, the Schools Council Project on Technology.
7. *Practical Work in School Science* (Leicester University Press, Leicester, 1963).
8. Nuffield Chemistry Introduction.

# 9 IMAGINATION AND TEACHING

The underlying theme of this book has been the essential part played by imaginative processes in the learning of science. From whatever aspect we examine laboratory teaching, be it the items of equipment, experimental design, the naming of concepts or the flights of pure theory, the analysis leads us every time to the same goal – the stimulation and preparation of imagination. There is no routine measurement or experiment so practical or humdrum that it can be carried out without such aid. All experiment is performed in imagination every bit as much as in tangible reality.

By this word 'imagination' I do not mean the weaving of a fantastic story, even though the way in which young children think about science often seems to be close to fantasy. I use it rather to include all the internal, personal ways of thinking about the objects of nature. Sometimes these ways will be purely visual in the manner of those excellent educational films which show, by animated diagrams, the pumping of the heart or the action of peristalsis. At other times the image to be evoked has been internalised much earlier by motor action or by the sense of touch, as was shown in Chapter 5. One is compelled to add that children also use a special kind of intense personal involvement where they project themselves wholeheartedly into the role of an animal or a molecule with surprising results! Particles in a heated solid vibrate more and more until they 'cannot stand it any longer and they melt'. The veins of leaves hold them open to the sunlight, 'hoping to be able to start photosynthesis'. Electric current leaps across a short-circuit and 'gets so excited that it melts the fuse'.

This attractive, childish manner of writing about science shows the young imagination at work in its own special way which is quite vital to the process of learning. It signals to the teacher how far mechanistic explanation has developed, but may not always be as primitively animistic as the words would suggest to an adult. Even in the accepted language of physics we speak of 'excited' electrons; the vivid everyday vocabulary of children is both valuable and inevitable for expressing their understanding of science.

We have seen how, in the 1960s, school practical work was exalted to the pinnacle of pure 'discovery', the quintessence of the whole scientific enterprise, so it seemed. Looking back from this vantage point it is easy to see that their experiment was not so much founded on prepared imagination as designed to stimulate such imaginings in sadly empty soil. If the method did succeed in the hands of some inspiring teachers then we should look to this inspiration, the source of imagination, as the instrument of success. Interestingly, those who have tried to prove the value of this discovery method by careful and objective research have met with little success. They had to avoid any aids to preconception, which is the stuff of imagination, and their results were correspondingly disappointing.[1] School science teaching owes a vast debt to these discovery enthusiasts for the emphasis that they laid upon practical work even if, as appears to be the case, they mistook the order in which imagination stands to the act of experiment.

Modern society is very sympathetic to the child's own vision of the world, encouraging wonder and inventiveness to a level which no longer just apes adult achievement. Only in the field of practical scientific work has the mistaken attempt been made to banish imagination in favour of disinterested observation. This belief in the neutrality of experiment is hard to dislodge; probably it reflects some longing for a standard of security among all the difficult adult value judgements that we are called upon to make. So we confer upon the 'scientific method' of experiment a certainty and an objectivity which it cannot possess and indeed rarely claims for itself.

The imaginative power that we want to encourage is not static either in its content or in its use. Every theory that we teach is concerned to mirror happenings and movements in the natural world and so the concepts on which they are based have to be refined and sharpened until they can be used with the precision of mathematics. Here is the peculiar challenge of science teaching. Imagination, concept-forming, must by its nature be an introverted private activity, and yet it is our task to read this inner picture that our pupils form, to help them shape and colour its vague contours by discussion and experiment so that they can correctly speak and understand the image-language of science. This means that we hope to cultivate the growth of these special kinds of thinking.

## How Does the Imagination Grow? Theoretical Suggestions

From a teacher's point of view most theories of learning are too descriptive, too passive, to be really helpful. Psychological research only aims to observe and classify the stages of development in children's thinking, whereas the teacher's task is to help their thinking progress and mature. What conjectures have been made in this field all too often ignore the accelerating effect of either teaching or indeed of any kind of human converse and interpretation. This is all the more extraordinary in view of the overwhelming evidence that a lack of human conversation and encouragement can be quite devastating to the mental growth of very young children. Nevertheless the following outline of theories is valuable because it explains some of the more curious suggestions in current educational innovation.[2]

(1) Piaget's early thinking on this subject simply supposed there to be a connection between mental and organic growth: 'the development of brain structures are, in some manner, isomorphic with the emergence of formal thought.'[3] In the same book the authors concede that in primitive societies children never do reach this stage of formal reasoning. It is most unlikely that the potentialities of their brain structures are any different from our own and yet the forces of adult stimulation or teaching found no place in this theory.

(2) Many others, from Rudolph Steiner to latter-day Piagetians, have suggested that it might be the inner conflicts brought about by social puberty, the struggle to carve out an individual place in adult society, which act as an incentive to move into new, more formal methods of thinking in which abstract probabilities can be handled in addition to concrete operationalism. No teacher would deny the reality of these strains and, for girls in particular, they can bring about an awkward lethargic moodiness which sometimes inhibits creative thought for as much as a whole year. There is little in classroom experience, however, to show that such pressures are either a sufficient or necessary condition for mental growth.

(3) Jerome Bruner took the argument a little further away from that of biological maturation by proposing an oyster-and-pearl model in which it was external irritation which triggered mental progress: 'the "operations" of which Piaget writes so brilliantly. . .grow in terms of the changing objectives. . .and are responses to the vicissitudes through which a person is going in pursuit of his objectives.'[4]

(4) One of the characteristics of Jean Piaget's theory of cognitive development is the series of well-known 'levels' – plateau regions in a child's mental capability – at which he postulates that they rest for years at a time without venturing up to the next level or slipping backwards to an earlier staging post. Such ideas make his theory cumbersomely static. He has always been most emphatic that children

can only learn matter which is appropriate to their particular cognitive level.

Growth is explained by the Piagetian theory of 'equilibration' which is brought about by a process called 'self-regulation'. This appears to be a kind of mental homeostasis whereby stimuli from the environment cause the child to adjust his internal cognitive balance, as it were, in order to cope with them. (How this squares with the previous point that children are unable to profit from information which is inappropriate to their current thinking mechanism is left totally unclear.)

It was perhaps inevitable that such an internal model of mental growth should lead, once again, to a personal discovery approach to learning. When Piaget turned his attention to school education he advocated 'the use of active methods which give broad scope to the spontaneous research of the adolescent and require that every new truth to be learnt be rediscovered'.[5] This was all the more surprising since followers of Piaget have often quoted investigations of children's learning which seemed to show that even practical evidence from the weighing of different shapes of plasticine did not convince young children of the conservation of weight in cases where they had not already reached this stage of thought.

(5) There is one other theory that deserves a mention because it is sometimes quoted in an educational context. This is the Personal Construct Theory of G. Kelly. Put simply, this cuts right across all the Piagetian sequences of thought and simply asserts that every individual understands happenings in his own way. Each of us keeps making and changing our mental constructs so that we can explain and anticipate events.

Although it was not originally conceived in this context, Construct Theory does sometimes match very closely with the partisan approach to experimental learning which was put forward in Chapter 7: 'it does broadly suggest that even the most obvious occurrences of everyday life might appear utterly transformed if we were inventive enough to construe them differently.'[6]

However, it is disappointing to find once again that the communication of mental pictures through personal contact finds no place in this theory. Kelly was at pains to point out that everyone does his own private 'construing', but when such an attitude is applied to education it can only lead to the kind of isolated independent learning that we encountered in Chapter 4. 'If the goal of education is seen as the elaboration of unique outlooks, then the number of different kinds of educational content is limitless and there can be no inherent superiority or inferiority as between them.'[7]

**Training the Imagination**

Where does all this leave the science teacher who hopes to train and teach the imaginations of his pupils? Despite the lack of prestigious theory to back his practice, there are still two very important areas on which he can draw that have been almost totally neglected by the forgoing theories.

In the first place the study of science cannot be equated with the cultivation of divergent personal creativity. However great the room we allow for invention and private research, however imaginative the fabric of scientific theory, it is not finally susceptible to individual interpretation because it has to be communicable knowledge. In the last analysis this is not a matter on which a student may adopt a personal stance and insist on his own viewpoint. Science is a social activity and its images are as clear, or clearer, than the vocabulary of any other language.

In the more difficult matter of a theory of teaching it would be well to revert to the simplest and most basic types of learning. The youngest children, like the young of all species, learn from their parents by the process of imitation. This takes place, it is true, in familiar surroundings and an atmosphere of affection and respect. What we attempt in the laboratory, although far more specialised, follows this same infantile pattern. It involves cultivating our pupils' thought by making it easy for them to understand our images and use our methods of reasoning. It includes showing and giving as well as trial and imitation – a commerce far more subtle than dogmatic instruction because it employs interpersonal communication and socialisation at a deep and valuable level.

When a child does not just imitate but learns through imitation the idea being assimilated becomes thoroughly internalised and mentally digested so that its substance can be used again in different circumstances for which it may seem appropriate. Here is a signal point of contrast between this kind of training of the imagination and simple instruction. Through continual contact, in sequences of lessons, through a period of months and years, the process continues, giving the teacher an opportunity to watch and help the internalised images develop in ever more precise and complex ways. Gradually the teachers see the seeds that they have sown and nurtured blossom as their pupils acquire the habit of abstract thought using mental analogies which combine taught methods with their own growing vocabulary of images.

Because imaginative teaching is a two-way method of communication, it is vital for the teacher to have formed some assessment of the cognitive methods of his pupils. In this context the Piagetian classification can be useful, although the practice of teaching delivers some surprising insights into the capabilities of children and occasional body-blows to orthodox theory. Experience does not confirm, in detail, either the age range of the cognitive sequences or their static quality. Pupils can be observed changing their methods of thought both upwards and downwards – this latter especially when beginning a new topic. Quite young children can emerge from a reverie to exhibit a piece of abstract reasoning which no psychologist would have credited at that age:

(1) A six-year-old boy, having watched children jumping off the top diving board together, started thinking about gravity and rates of falling. The picture in his mind was so vivid that he was almost laughing as he explained to me that 'if a man with a suitcase was falling he could leave go of the suitcase, it would stay beside him and he could think that gravity had all gone!'

(2) A nine-year-old girl was in a class of pupils making weighing machines. To obtain greater accuracy they were finding places to set a 1g rider so that they could weigh ¼g, ½g and ¾g. The girl grew worried by her thoughts: 'Miss, I know this is a silly question, but. . .the moving bit still weighs 1g wherever we put it. . .how can it be ½g in one place and ¼g in another?'

Admittedly such cases are unusual. Class teaching must necessarily be aimed at the more general level of imagination and failures to get across a useful method of thinking about an experiment demonstrate a lack of competence in the whole class. Consider, for example, an experiment in which an empty flask fitted with a long tube is inverted into a beaker of water and is first warmed with the hands and then left to cool. When the author first taught this lesson to first-year pupils in a secondary school it was discouraging to hear the following comments:

'the water is bubbling' – instead of the air is expanding,
'the water is expanding' – instead of the air is contracting.

Figure 9.1

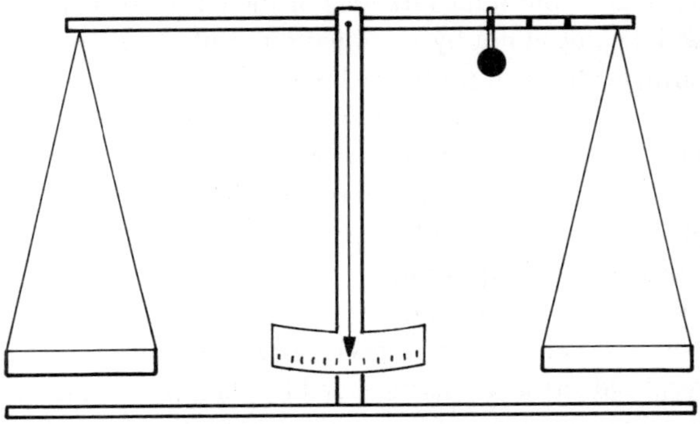

Owing to a shortage of science staff physics then had to be dropped for several years from the first-year time-table and it was found that the second-year pupils were better able to draw conclusions from this experiment than had been the case previously. Did this mean that a Piagetian barrier made this subject-matter unsuitable for the younger children? Re-teaching the experiment later to another first year proved successful only when prior training of the imagination had paved the way. There had to be not only some preliminary 'experience-type' experiments to show the reality and weight of air, but they also discussed what might happen if the invisible air got too big for the flask, or if it shrank and got too small. (Apparently they knew some fairy stories about people shrinking inside their clothes which helped a lot!) After this their written accounts of what had happened in the heating and cooling experiment showed almost 100 per cent success.

It might be expected that all 16-year-old pupils would be sufficiently advanced to be able to use operational thought quite spontaneously on the working of a switch in a circuit. Unfortunately this may not always be the case. A low-ability pupil who fails to forecast the effect of such a switch can make a dramatic leap forward if a tale of 'gates' and 'crowds' stimulates the sluggish imagination. This may take a personal effort at vivid communication with gestures and jokes to match; there is no mistaking the moment of internalisation of the image. Not just the circuit in question but diagrams of many more become delightfully clear and comprehensible by means of the imagined mechanism.

## The End of the Trail

By the time our pupils reach the sixth form we will have had a chance to tell whether careful and continuous training of the imagination has borne fruit. This will be judged not by examination results, nor even by their ease of learning, although it is to be hoped that both of these will have profited in great measure. If the training has been successful, they will now be able to generate their own scientific images independently of our promptings. These internal models will demonstrate originality by bearing the stamp of the pupil's own personality and aptitude as they are applied to a scientific situation.

The more practical pupil, for example, may be extending the practical work set on a certain topic in a special direction. I am reminded of a 16-year-old girl who was intrigued by thermionic emission to the extent of spending several lunch-hours timing the discharge of a gold leaf electroscope at different distances from a red-hot wire. The experiment was simple enough, but while she worked she spoke of a film she had seen of solar flares in which matter ejected from the sun rose to different heights in parabolic trajectories before falling back in the intense gravitational field. To her mind's eye the emitted electrons were as perceptible as the white hot plasma from the sun's interior, and hence the simple experiment had a vividness far beyond its commonplace appearance.

Other promising pupils may already be showing a more theoretical bent, and here too they will be exhibiting not only mathematical competence but also the results of imaginative training. This can be an impressive achievement. A 15-year-old boy invented his own proof that the mean free path of a gas molecule was independent of temperature by means of a thought experiment. Imagine, he said, a film of the kinetic activity of gas molecules, an increase of temperature could be represented by speeding up the same film where there would be no change in the distances travelled between collisions. This use of scientific visualisation showed promise of a high order.

At this low-water mark in educational thinking when the craft of teaching occupies such a neglected niche in the public estimation, it might be well to end with an uncompromising affirmation of its values and aspirations:

> a teacher. . .first provides recognition for his students. He recognises what they bring to their education and he lets it speak. He gives it speech. Then he makes a line between that experience and the

possibility of other experience. He offers the pupils the recovery of their lost chances to seize the life of their imagination and make it command the vacancy of the world. He helps to turn images into theories.[8]

## Notes

1. There is a very interesting piece of unpublished research by T. Kyle (Glasgow, 1950, quoted in SCSB 3 Evans Methuen, 1970) concerned with a group of 14-year-olds left to discover Archimedes' Principle on their own. Observation showed that even finding the volume of the immersed object was performed in reverse – submerging it first – as if by foresight. Since none had the foresight to look for a quantitative upthrust, the Principle was not discovered.

2. *Alternatives for Science Education. A consultative document* (ASE, Hatfield, 1979). Independent Learning in Years 16-18.

3. J. Piaget and B. Inhelder, *The Growth of Logical Thinking: From Childhood to Adolescence* (Routledge and Kegan Paul, London, 1958).

4. Review article in the *British Journal of Psychology*, no. 50 (1959).

5. J. Piaget, 'To Understand is to Invent', *The Future of Education* (Penguin, Harmondsworth, 1976).

6. G. Kelly in D. Bannister (ed.), *Perspectives in Personal Construct Theory* (Academic Press, New York, 1970).

7. P. Salmon and D. Bannister, 'Education in the Light of Personal Construct Theory', *Education for Teaching* (1974).

8. F. Inglis, *Ideology and the Imagination* (Cambridge University Press, Cambridge, 1975).

# INDEX